Steam Microturbines in Distributed Cogeneration

Jan Kiciński · Grzegorz Żywica

Steam Microturbines in Distributed Cogeneration

Springer

Jan Kiciński
Grzegorz Żywica
Institute of Fluid-Flow Machinery
Polish Academy of Sciences
Gdansk
Poland

ISBN 978-3-319-12017-1 ISBN 978-3-319-12018-8 (eBook)
DOI 10.1007/978-3-319-12018-8

Library of Congress Control Number: 2014951670

Springer Cham Heidelberg New York Dordrecht London

Printed on acid-free paper

Springer is part of Springer Science+Business Media (www.springer.com)

Preface

The analysis of the dynamic properties of fluid-flow machines has been for many years the subject of numerous studies conducted in many centres around the world, both academic and industrial. The results of these studies contribute to the development of theory, but also have direct application use and often end in implementation. In recent times, works on the improvement of rotating machines and a better understanding of the phenomena taking place in them, are growing in importance. Dynamic problems have escalated, as global competition forces manufacturers to continuously improve the operating conditions of new products, while maintaining a light, and more and more frequently even miniature design. In addition, raising the commercial characteristics and miniaturization is occurring along with the growing requirements in terms of efficiency, reliability and precision of operation. Due to the fact that fluid-flow machines are used in many sectors of the economy (among others: energy, transport, aviation) and very often constitute a key element in larger installations, the growing expectations also apply to such machines.

The study of the dynamic properties of various types of fluid-flow machines has been conducted for many years, also at the Institute of Fluid-Flow Machinery PASci in Gdansk. In previous years, the works were focused mainly on large machines such as turbine sets used in the power industry. Although microturbines are machines that are much smaller than thermal turbines used in the commercial power industry, very complex dynamic problems also occur in them. Due to the smaller size and greater rigidity of rotor blades, most problems associated with vibrations of these elements can be avoided, however, other difficulties appear, which do not occur in large turbine sets, for instance, high rotation speeds, aggressive working mediums and technological difficulties in making very small elements. As a result of the fact that microturbines are widely used in distributed generation, specialized diagnosticians rarely keep watch over their proper functioning, which is standard in the case of professional power turbines. Therefore, microturbines should also be reliable and work safely. It is additionally very important to use appropriate numerical models of microturbines, enabling their analysis and optimisation already at an early stage of design works.

In the era of constant development of computer technology and measurement methods, this is taking on a particular importance. By using the calculation tools available, the modern engineer is able to model and do simulations of even very complex mechanical systems. Simulation tests of a well-identified machine model can provide a lot of valuable information. A properly developed and used model can contribute to obtaining the desired effects much more quickly and cutting down the scope of more expensive experimental studies of prototype designs to a minimum. The described problems justify the need to carry out research for a better understanding of the phenomena occurring in different types of fluid-flow machines, including steam microturbines. This monograph has been devoted precisely to those issues.

The following chapters discuss simulation testing and experimental dynamic properties of microturbines, as well as their subassemblies. Chapter 1 provides the introduction to renewable energy and distributed generation, indicating the possibilities of using microturbines in the economy. Chapter 2 characterises the microturbines and discusses the key problems concerning the properties of dynamic mechanical systems, in particular rotating systems. It also provides information on the research tools used. Chapter 3 presents the results of studies of unconventional bearing systems used, among others, in steam microturbines. Foil bearings were focused on especially. Then, in Chap. 4, methods of modelling and simulation tests of microturbines are discussed, on the example of two alternative structural solutions—radial and axial microturbines. Experimental tests of the properties of dynamic elements of microturbines are discussed in the last, fifth chapter.

The published monograph should in particular interest the group of people who work in designing, modelling as well as operation and diagnostics of various types of fluid-flow machines, especially steam microturbines. The authors of the work are hoping that this position will contribute at least a little to a wider application of modern research tools in the design, operation and diagnostics of such machines, which in turn will also contribute to introducing more elaborate and reliable constructions on the market.

The results presented in this monograph are the output of several years' study on the development of simulation models as well as structural and dynamic analysis of the microturbines. The research was conducted within the project POIG.01.01.02-00-016/08 "Model agroenergy complexes as an example of distributed cogeneration based on local renewable energy sources" co-financed by the European Union, and also within the project "Advanced Power Generation Technologies—Task 4. Development of integrated technologies for the production of fuel and energy from biomass, agricultural waste and other", from the National Centre for Research and Development. The most important effect of the realization of these projects is, inter alia, the creation of prototype small power plants with ORC (owned by IMP PAN), with different power outputs as well as several construction solutions of steam microturbines.

The book was written by two authors, but its content was influenced by the work of many people. At this point the authors would like to thank their immediate colleagues from the Department of Turbine Dynamics and Diagnostics

IMP PAN, in particular M. Bogulicz, M.Sc. for the involvement in the creation of computer programs, P. Baginski, M.Sc. Eng. for cooperation in the conduct of experimental research and B. Czoska, M.Sc. for his assistance in preparing the translation. During the realization of the research also participated employees of the Aeroelasticity Department IMP PAN in persons professor R. Rządkowski, Ph.D., D.Sc. and M. Drewczyński, Ph.D. and staff from other research centers with which the authors have been cooperating with for many years, including a team of professor Z. Kozanecki, Ph.D., D.Sc. from the Lodz University of Technology and also the team from the Faculty of Technical Sciences, University of Warmia and Mazury in Olsztyn, directed by W. Miaskowski, Ph.D. The authors express their thanks to all members of the teams involved.

Contents

Chapter 1
Distributed Cogeneration. Civic Power Engineering. New Ideas

Abstract This chapter provides an introduction to renewable energy and distributed cogeneration. It discusses the European Union's ongoing policy and plans in the area of energy. Particular attention was drawn to prosumer energy generation. The terms: smart grid and energy-plus technologies were clarified and their advantages were presented. The installation examples in micro and small scale are given, such as the micro CHP power plants with ORC, having different power outputs. Their capacity to cooperate with other eco-energy installations were also demonstrated. It also presents various alternative microturbine constructions that can be used in ORC installations contributing to electricity generation. The next section discusses the concept and examples of autonomous energy regions (ARE), which have the potential to revolutionize power industry. As an example of a large plus-energy building, which will be used for various new technologies of heat and electric energy generation from renewable sources, the new research center of the Polish Academy of Sciences in Jabłonna near Warsaw was presented.

1.1 New Energy Policy of the European Union. Civic Energy Generation

Following the recent actions of the European Union, it can be concluded that the previously approved strategic direction of the EU policy aimed at reduction of emissions, introduction of renewable energy sources and energy conservation in buildings will not only continue, but also will be strengthened. Framework activities in this area are defined by **Energy Road Map 2050 (ERM2050)**, which is currently under preparation, new support mechanisms in the form of Framework Programme "**Horizon 2020**" (research and innovations) as well as by already functioning tools such as "**SET Plan**" (implementations, cooperation with industry), which are currently being improved and broadened.

One thing is certain: the above-mentioned EU legislation clearly determine the path for the energy system development based on renewable energy sources, in particular small-scale and distributed.

J. Kiciński and G. Żywica, *Steam Microturbines in Distributed Cogeneration*,
DOI 10.1007/978-3-319-12018-8_1

In the light of the comments on these legislative initiatives, regardless of the pace of the works and the dangers arising from the lack of a stable law, one can expect a dynamic development of **intelligent and distributed energy systems** as well as significant improvement in energy efficiency. In this context, it is important to stimulate the development of activity of **prosumers** (producers who are simultaneously the consumers of energy), i.e. millions of small investors who are micro-scale producers of energy for their own use, selling the excess energy produced to the national power grid. This idea is a great opportunity for the traditional large-scale power engineering, because the possible lack of capacity cannot be so quickly replaced by large conventional units (new power units) entailing considerable investments.

Additionally, this book discusses the conditions and opportunities for the development of prosumer energy generation or more broadly: civic energy generation in Poland and other countries. **Civic energy generation** is a great vision in which the citizen becomes an entity and not the subject of the energy market, and additionally has its virtual advisor in the form of smart grid and data processing technologies in a "digital cloud".

Studies on the conversion of energy from renewable sources require a multidisciplinary approach. Hence the need for a systematic synergy of scientific achievements in the field of technical, biological and agricultural sciences, but first of all in the field of ecological power engineering.

Summarizing the above considerations it can be concluded that the distributed energy engineering based on renewable energy sources (DES/RES), "Smart Grid" and applications in the "digital cloud" will be a key element of national energy policy in EU in the near future. This means that the hybrid systems integrating solar, biomass and wind technologies, heat pumps and energy storage for energy-plus residential houses, facilities and housing estates co-working in a smart grid are a challenge for the moment. In this context, the main task for national research centers, industry and self-government entities will be to develop economically viable solutions for the market using multivariate synergies between the different technologies and the elaboration of the most useful applications for the processing in "digital cloud" [1–10].

The development of distributed energy generation system DES is directly related to the implementation of **intelligent energy management systems** (SG—Smart Grid). Smart grids will enable easier connection of distributed sources to the National Energy System, reduce the grid load and minimize the risk of blackout.

It is worth to emphasize that a smart grid itself in the electricity distribution system is only a certain technical layer. More important is that through the use of this grid, one can generate new services as well as new applications and as a result—simply make money. The grid itself is not able to trigger such changes but attractive applications are. It is worth noting that, for example Apple company succeeded predominantly thanks to applications rather than smartphones themselves.

In order to take advantage of all its benefits, the smart grid must be provided with appropriate **IT interface layer** enabling prosumers' communication in the so-called "cloud computing"—Figs. 1.1 and 1.2. The term "cloud computing"

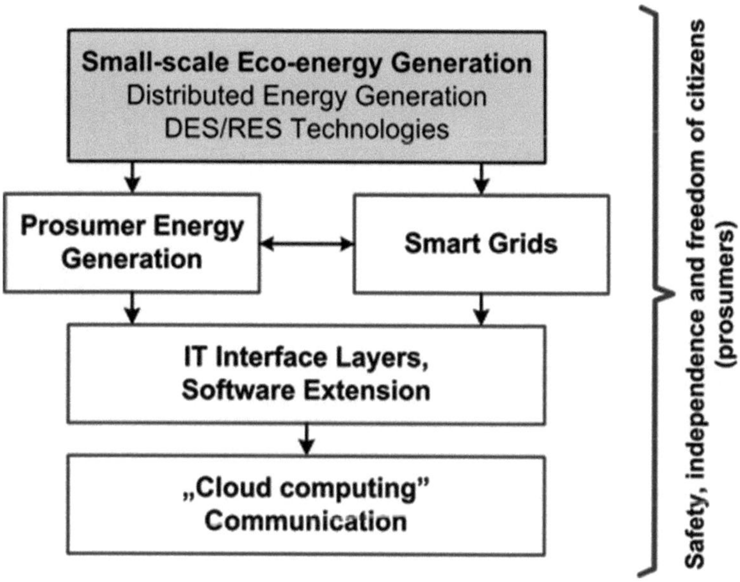

Fig. 1.1 The development of small-scale eco-energy production as a vision of civic energy generation: from DES/RES technologies to processing in the "cloud". The vision of safety, independence and energy related freedom of the citizens

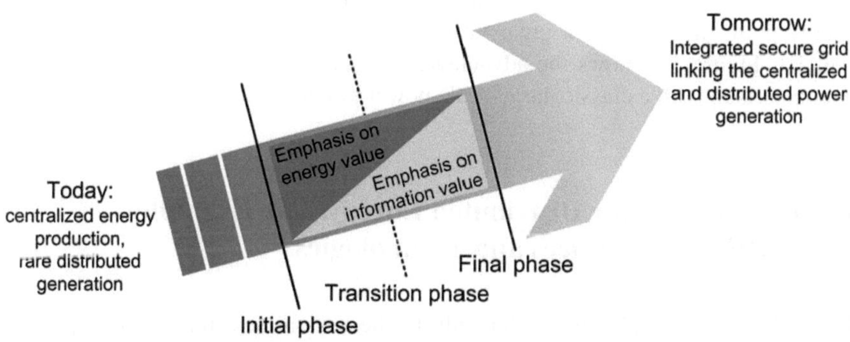

Fig. 1.2 The role of information in the development of energy systems

is here quite contractual. Although it is used in the literature in rather different contexts, its main features will find application in relation to the distributed eco-energy generation. There will be a place for different kinds of virtual installations, business models or even advisory systems applying artificial intelligence methods. The "cloud", defined and understood as above mentioned, will be in fact **a virtual platform for the exchange of information for prosumers**, their data and knowledge bank as well as their system of professional advice. This is undoubtedly the future of this sector of energy engineering and a targeted vision of **civic energy generation**, where prosumers and consumers feel independent and free.

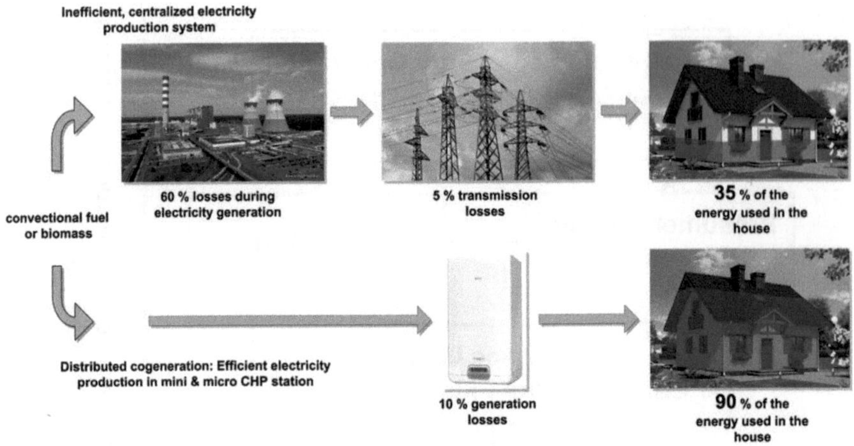

Fig. 1.3 The advantages of distributed cogeneration in comparison with centralized power generation

Similarly to the rapid development of ICT, which resulted in freedom of communication among citizens, one can expect that the development of distributed technologies (DES/RES) in combination with the smart grids and virtual communication platform for prosumers will make the citizens free also from the energy point of view. It's a great vision, which shall be given more attention—Fig. 1.1. Figure 1.3 shows the advantages of small-scale distributed cogeneration in comparison with a classic, large-scale power system.

1.2 Installations for Distributed Energy. The Examples of DES/RES Devices and Technologies

The following examples will refer only to the several selected research projects coordinated by the IMP PAN (Institute of Fluid-Flow Machinery of the Polish Academy of Sciences) in Gdańsk. However, these are the largest research projects in the country regarding the field of RES-based eco-energetics. They can therefore be treated as representative ones for this energy sector in one of the EU countries (Poland).

This chapter will focus on the results of the work related to so-called **"energy-plus"** technologies and **small and micro-combined heat and power units**. These results are the effect of research conducted at the institute and in cooperation with industrial partners (mainly the Capital Group ENERGA) and more than a dozen research teams from different research centers across the country. These study focuses on the future implementation and is addressed to individual and municipal consumers.

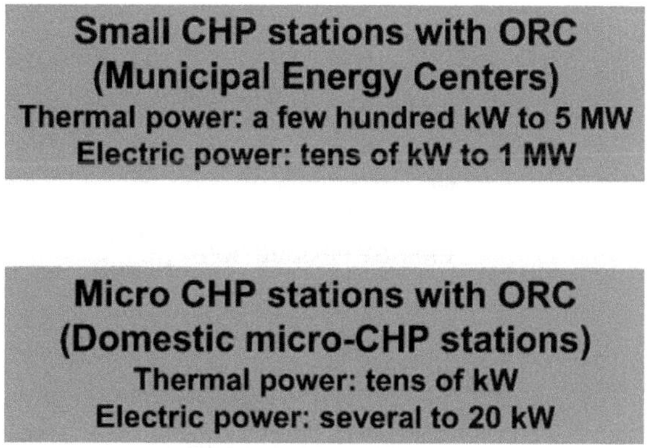

Fig. 1.4 CHP ORC stations in small and micro scale as suggested in the framework of the IMP PAN projects

It is not the intention of the authors of this book to present any other DES/RES technologies known in the country and especially in the EU even in the field of combined heat and power units.

When it comes to the ongoing studies, the construction of the CHP ORC plant is planned (the blocks consisting of a boiler and a microturbine operating with a low-boiling agent using an ORC) with the electric power of several kW and tens of kW of thermal power. In the framework of the another project it is planned to build CHP ORC units of higher power (hundreds of kW of electrical power, thermal power up to several MW).

The results of these projects will be thus addressed to individual customers in the form of **domestic CHP units**, and to the municipal customer as **Municipal Energy Centers** or **Autonomous Energy Regions (ARE)**. Figure 1.4 shows this pictorially.

1.3 Domestic CHP ORC Units

One of the main purposes of the projects is to elaborate several prototypes of the so-called domestic CHP stations, which will include biomass and multi-fuel boilers with a capacity of tens of kW coupled with micro-turbine with a capacity of a few kW—operating with a low-boiling agents using an Organic Rankine Cycle (ORC). The concept of Domestic Cogeneration Units i.e. small plants that produce heat and electricity, may be attractive to thousands of individual customers, especially if the micro-CHP units will be integrated with other systems (solar/wind/water energy, heat pumps and energy storage), creating the so-called **hybrid systems** with greater mutual synergy—Fig. 1.5. Such micro-CHP units could create a new and powerful market and play a key role in the concept of civic and prosument energy generation.

The IMP PAN group (that means: IMP PAN, TU Łódź, TU Gdańsk, University of Warmia and Mazury) developed two concepts of micro-turbines with a capacity

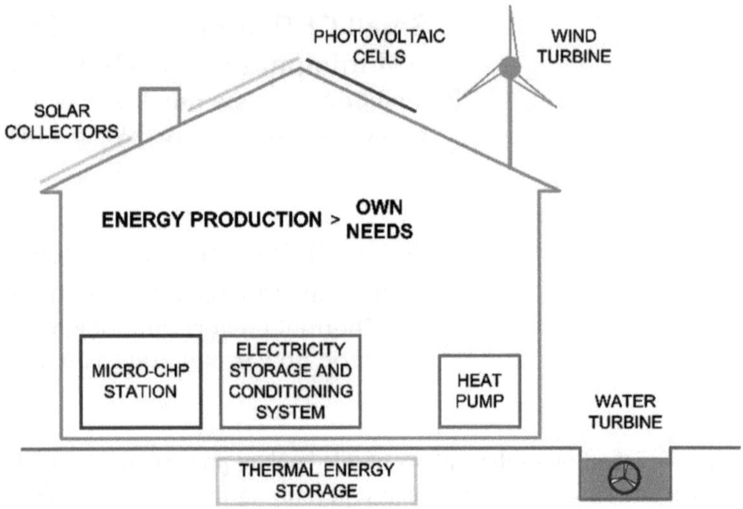

Fig. 1.5 Domestic CHP units—an example of the hybrid system with a much greater mutual synergy. The attractive vision of civic or prosumer energy generation for thousands of individual customers

Fig. 1.6 The possibility of utilization of a microturbine's working medium in the liquid and gas phases as a bearing lubricant. The concept of hermetic construction of a turbine and a generator to facilitate the integration with a boiler

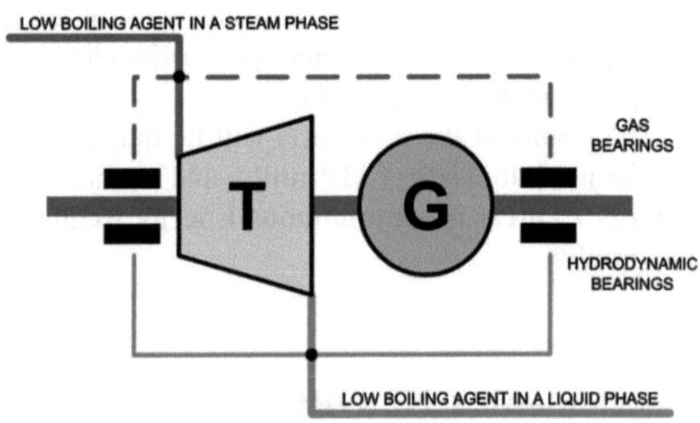

of 3 kW (axial-flow and radial-flow) coupled with multi-fuel boiler with a capacity of 20 kW (biomass or gas fired). As far as microturbines are concerned, the essential idea was to use the low-boiling agent (turbine's working medium) for bearing lubrication, which ensures tight and hermetic construction. Figure 1.6 shows this idea, while drawings and photographs of micro-turbines are presented in Figs. 1.7 and 1.8. Figure 1.9 shows the photo of test stand in the micro-turbine laboratory (located at the IMP PAN in Gdańsk) and the photo of multi-fuel boiler.

Currently, laboratory investigations are carried out and as a result the boiler and both versions of micro-turbines are being tested. After completion of tests, the development of a target version (and perhaps commercial one) of an entire micro-CHP unit is planned. The brief foredesign and initial documentation of such micro-CHP station have already been elaborated—Fig. 1.10. If these undertakings are successful, it will be **the first in the country construction** of this type.

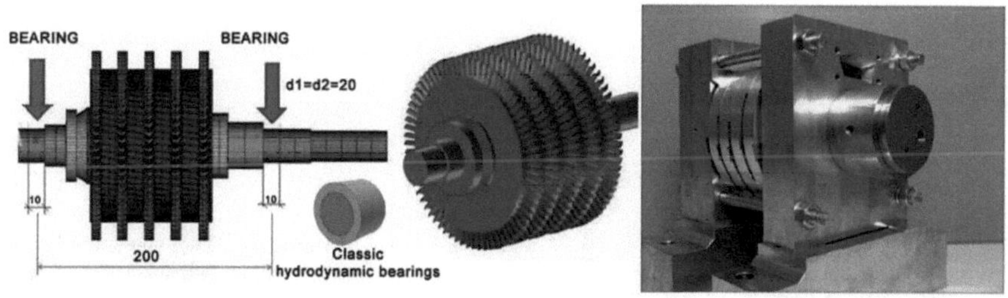

Fig. 1.7 The first concepts: the axial-flow micro-turbine, 5 stages, with a capacity of 3 kW and rotational speed 8,000 rpm. Tests are currently underway in the laboratory of the IMP PAN (developed by a research group of the Technical University of Gdańsk)

Fig. 1.8 The first concepts: radial-flow micro-turbine, 4 stages, with a capacity of 3 kW and rotational speed 24,000 rpm. Tests are underway in the laboratory of the IMP PAN in Gdańsk (developed by a research group of the Technical University of Lodz)

Fig. 1.9 The laboratory of the IMP PAN in Gdańsk: pictures of micro-turbine test stand (*on the left*) and multi-fuel boiler (*on the right*)

Figure 1.11 shows the developed targeted versions of microturbines and Fig.1.12 working version of wind turbine of low power (2–3 kW).

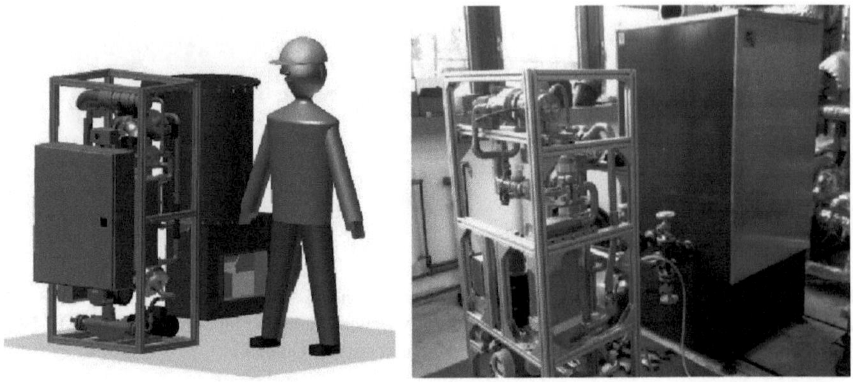

Fig. 1.10 Target design of domestic micro-CHP unit with ORC after all examinations and tests in the laboratory of the IMP PAN in Gdansk. *Left* the first concepts, *right* commercial version of the installation

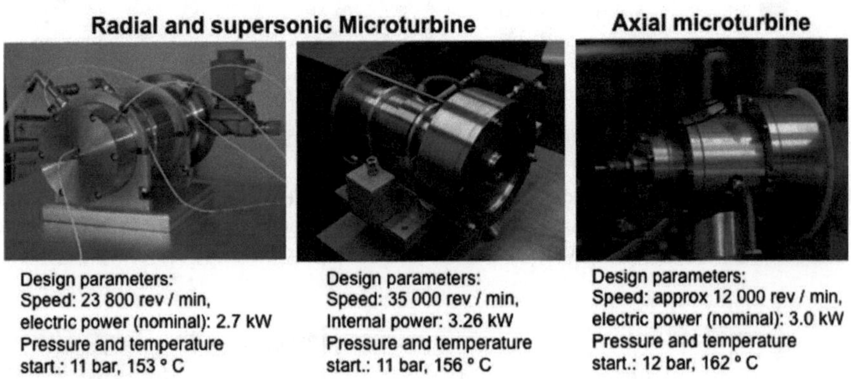

Fig. 1.11 Microturbines dedicated to CHP ORC units. Elaborated in TU Łódź and TU Gdańsk in cooperation with IMP PAN

Fig. 1.12 Two-rotors wind turbine with a *horizontal axis*. The proposed design solution enables the use of high-speed wind turbine power generator without the use of multipliers. Elaborated in TU Warmia and Mazury in Olsztyn in cooperation with IMP PAN

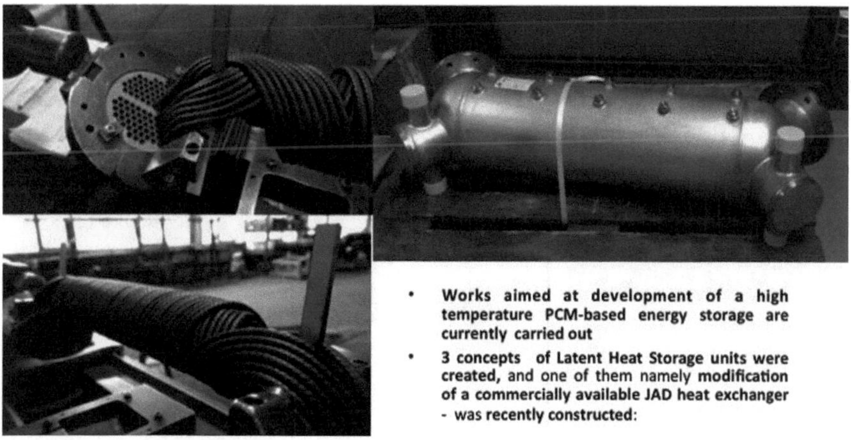

Fig. 1.13 PCM-based energy storage prototype in the lab in the IMP PAN (testing rig is currently under construction)

An important element of the research are also energy storage devices. In the course of construction are modern stations with PCM materials that complement the concept of home power plant—Figs. 1.13 and 1.14.

As part of the conducted projects, an attempt was made to develop a "software upgrade" for the Smart Grid systems. These are the origins of an advisory system which cooperates with an individual user in the virtual space. The system, called **SoftRol** was developed by a group of the Faculty of Technical Sciences, University of Warmia and Mazury in Olsztyn—Fig. 1.15. Currently the database (data on technology, type of crops, the size of an acreage) and also the knowledge base (simple business models) are being built along with testing of the entire system.

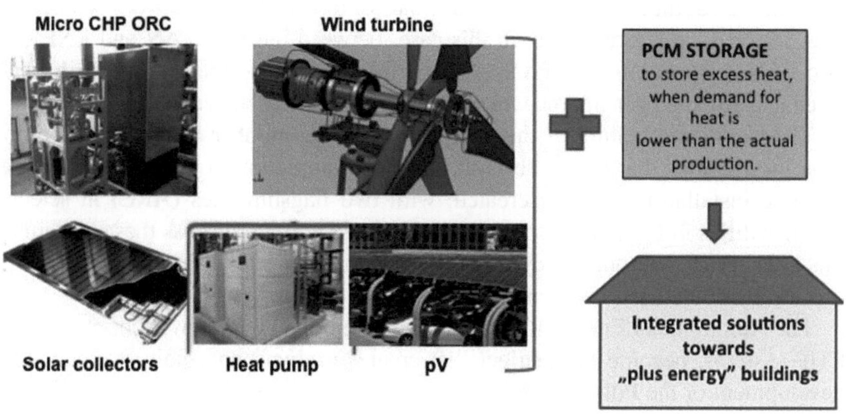

Fig. 1.14 Energy storage—a key factor in plus-energy buildings

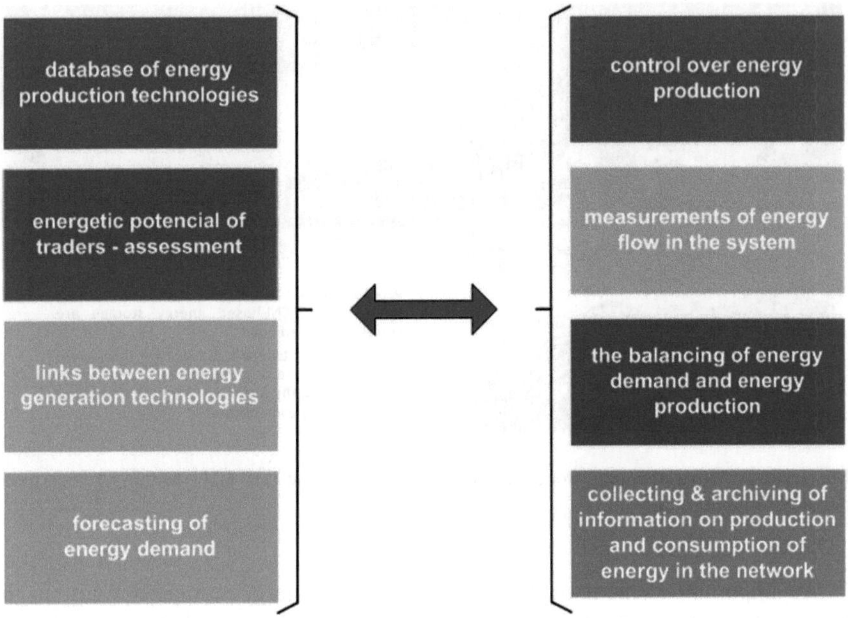

Fig. 1.15 SoftRol system as an example of data and knowledge base ("software upgrade" for the smart grid systems). SoftRol functions (*on the left*), smart grid functions (*on the right*). Developed by a group from the faculty of Technical Sciences, University of Warmia and Mazury, in Olsztyn. The system currently undergoes a test phase

1.4 Autonomous Energy Regions (ARE)

Concept, assumptions and building of the Autonomous Energy Regions (ARE) or Municipal Energy Center is one of the main tasks of the conducted projects. Construction of the ARE is an excellent example of cooperation between science and industry on such a large scale, in this case, between the IMP PAN and ENERGA Group. In cooperation IMP PAN—Energa the modern installations are built, dedicated to distributed generation, which are demonstratively shown in Fig. 1.16.

The significant feature of this cooperation is the number of innovative, prototype installations that are to be built upon its completion in 2015. Over a dozen of these installations will be created, with two flagship ones (ARE) in selected municipalities in Poland. For obvious reasons we will not discuss these systems in this book. ARE will play a key role (along with Domestic CHP Units) in the concept of distributed and civic energy generation.

The Autonomous Energy Regions (ARE), equipped with new technologies and CHP systems, become the **project** that creates a great opportunity for economic development of the Polish countryside. ARE is a concept that should direct Poland into the path of changes that are already occurring around the world. Currently, in Europe and in North America, as well as in China "green" villages, housing

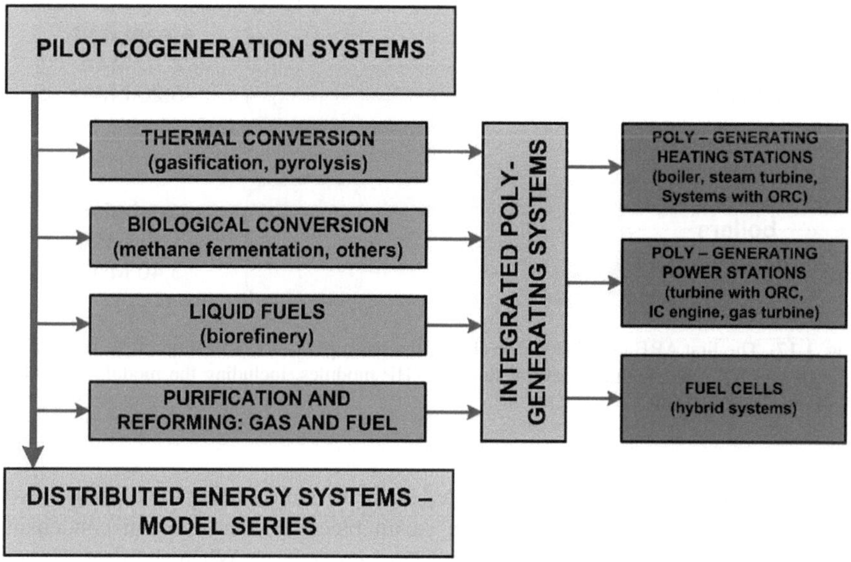

Fig. 1.16 Installations built in the framework of the IMP PAN and Energa group dedicated to the ARE and distributed generation

estates, districts and even cities are being developed (where "green" means emission-free, zero-energy or energy-plus).

The installation which is currently being built in one of the communities in Poland will serve as a **role model of a small-scale CHP plant**. In place of the outdated heat only generating, modern cogeneration units will be built, which include—Fig. 1.17:

- ORC system with an electric capacity of 0.143 MWe and the heating capacity of 0.83 MWt,
- CHP system with two combustion engines powered by natural gas with a total electric capacity of 3.1 MWe and the heat capacity of 3.4 MWt,
- Biomass-fired steam block with the electric capacity of 2.7 MWe and heating capacity of 5.2 MWt,
- Supplemented with an upgraded exhaust gas extraction system and backup boilers.

The solution shown in Fig. 1.17 exhibits the unquestionable technological progress in the field of energy solutions in the Polish municipalities.

What is the essence of the proposed first ARE in Poland? The biomass CHP system with ORC—designed to cover the summer demand for domestic hot water—will be working continuously throughout the whole year, while other systems will be started in response to the heat demand (this limitation is particularly important for combustion engines, for which the income from electricity production does not cover operating costs and is balanced by the income from heat generation).

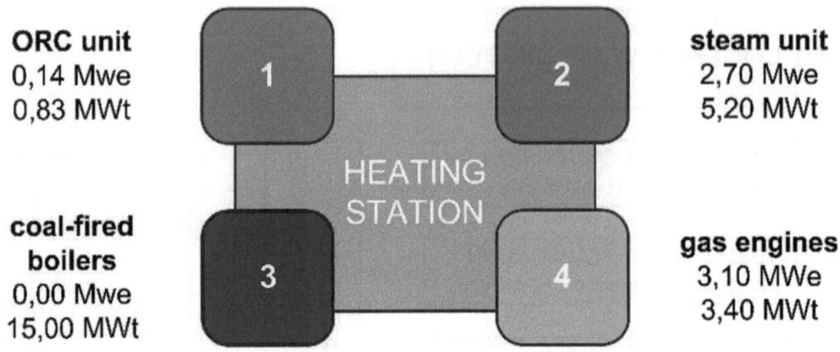

Fig. 1.17 The first ARE in Poland—the clue of the concept of modernization of the heating station in selected municipalities: several flexible CHP modules, including the module with ORC working all year round

The steam block is planned to be operated all year round. During the heating season it will work in a heating mode (with steam bleeding), and then in condensing mode with maximum efficiency in electricity production. When the heat demand increases then reserve, coal-fired boilers will be switched on.

The above idea solves an extremely difficult problem of this type of heating stations in the country i.e. what to do with excess heat during summer?

It is estimated that in Poland there exist over 300 heating stations of a similar, outdated type! If the idea is successful (and success is highly probable due to the scale of involvement of a large industrial partner), suggested solutions can be copied across the country in municipalities and housing estates.

Laboratory of the micro steam power plants is currently constructed in the IMP PAN in Gdańsk—Figs. 1.18 and 1.19. This modern and largest laboratory in the country is financed from the EU funds. In the future, this laboratory will also serve as a back-up research facility for devices and technologies to be applied in ARE.

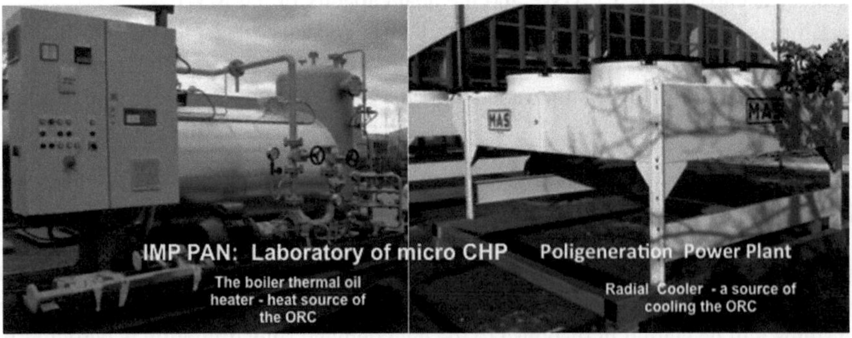

Fig. 1.18 The largest in Poland micro CHP Laboratory: the poligeneration power plant in the IMP PAN in Gdańsk

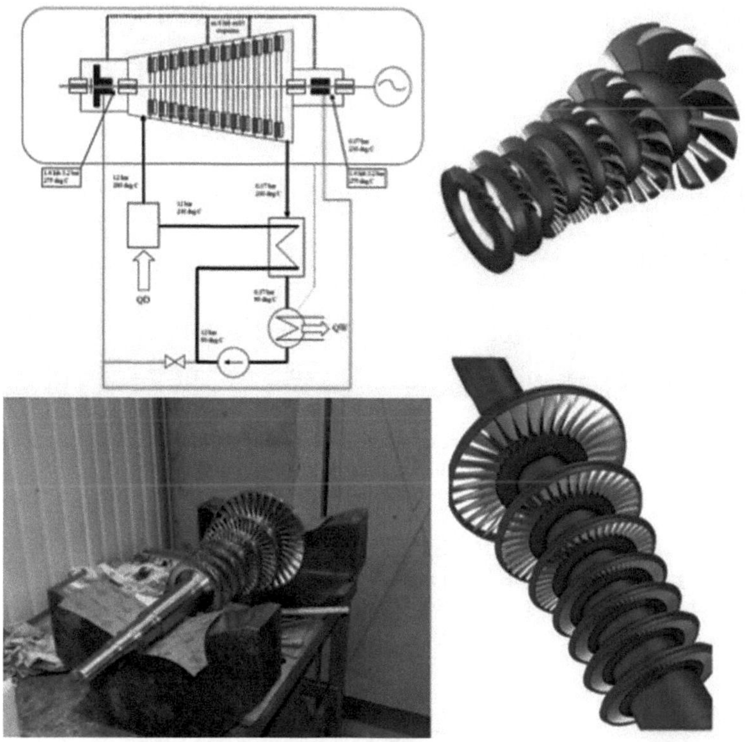

Fig. 1.19 IMP PAN Laboratory: ORC turbogenerator (own solutions) for ARE. Axial turbine 9,000 rev/min, 100 kW

1.5 Research Center in Jabłonna

Trends related to the development of distributed energy generation were the basis of decision made by the Polish Academy of Sciences and the Office of the Marshal of Mazowieckie Voivodship to build, a new and unique in the country, Research Center—Figs. 1.20, 1.21 and 1.22. This Center will be dedicated to the investigation of aspects of renewable energy conversion. The Center, (to be located in the municipality of Jabłonna, about 20 km from Warsaw), is planned to serve as a complex of modern research/testing laboratories with a unique apparatus and a set of demonstration facilities concerning distributed energy systems (DES) as well as technologies for the so-called energy-plus houses and facilities—Fig. 1.22. The beneficiary is the IMP PAN in Gdańsk, who developed the concept of the Center and coordinates its implementation. The completion of the investment is scheduled for December 2014 but activities related to design and legislative work, planning of research and consolidation of scientific and industrial centers are ongoing since 2012.

Fig. 1.20 The new "Energy-plus" Research Centre of the Polish Academy of Sciences in Jabłonna. Energy-plus technology for small private and public buildings

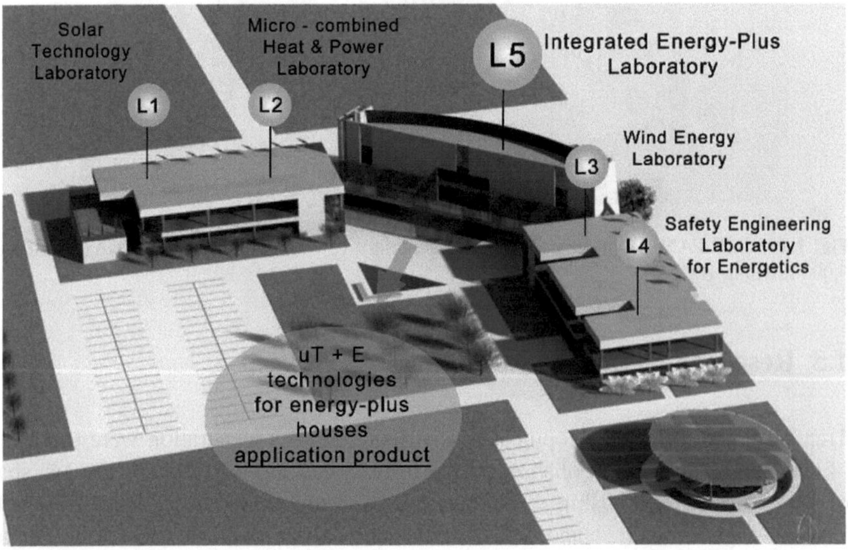

Fig. 1.21 The view of the Research Center in Jabłonna from the north—location of laboratories

The Center will be equipped with a local intelligent "Smart Grid" system which makes it possible not only to manage energy sources and the way energy is stored, but also remote archiving, transmission and processing of the experimental data obtained from research laboratories. The set of conference rooms, guest rooms, recreational areas and "RES" Education Path will enable Center to fulfill also the important educational and training functions.

The energy systems installed at the Center—in the power range from a few to several hundred kW (biomass and gas boilers, combustion engines, gas and wind

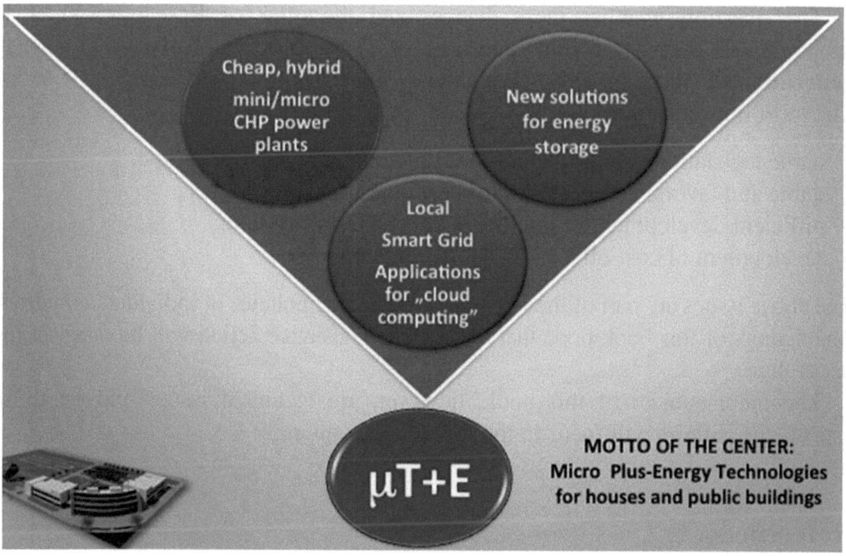

Fig. 1.22 The Research Center's offer for local scientific community, self-government and industrial entities. "Business card of the center"—Energy-plus technologies for houses, facilities and housing estates (uT + E)

turbines, heat pumps, solar collectors, PV cells and several types of energy storage), thanks to the possibility to create multiple variants and operational modules as well as the original hybrid associations, offer unique research opportunities for testing and certification of these devices. This is particularly important given the fact that all these devices are concentrated in one place and work within a single system of energy/information management (expanded BMS). The achievement of the so-called "synergy effect" in these conditions is certainly facilitated.

1.6 Concluding Remarks

This book presents only a few selected examples of specific devices (DES/RES) developed within research projects conducted by IMP PAN in Gdańsk. These are:

- Domestic CHP Units,
- CHP Plant in selected municipalities as an example of a ARE (Autonomous Energy Regions),
- Research Centre in Jabłonna—Energy-plus Technologies for houses, facilities and housing estates (uT + E).

All of the above installations and laboratories might play an important role in the development of small-scale distributed power generation in Poland and EU countries.

The question whether these devices and the offer of Research Center in Jabłonna will become an important element of the concept of **prosumer energy generation** or (more broadly) **civic energy generation** depends on many factors. The most important are:

- stable legislation supporting concepts and installations of this type;
- stable and favorable legislation for consumers and companies;
- sufficient development of intelligent “Smart Grid” system;
- development of cost-effective DES/RES technologies.

The above issues are part of the energy and economic policies of individual countries. The authors of this book hope that appropriate legislative action will be taken in the next future.

The main subject of this book, however, are technical issues and on those aspects the authors will focus in the following chapters.

References

1. Kiciński, J.: Do we have a chance for small-scale energy generation? The examples of technologies and devices for distributed energy systems in micro and small scale in Poland. Bull. Pol. Acad. Sci. Tech. Sci. **61**(4) (2013)
2. Yanine, F.F., Sauma, E.E.: Review of grid-tie micro-generation systems without energy storage: towards a new approach to sustainable hybrid energy systems linked to energy efficiency. Renew. Sustain. Energy Rev. **26**, 60–95 (2013)
3. Kaygusuz, A., Keles, C., Alagoz, B.B., Karabiber, A.: Renewable energy integration to smart cities. Energy Build. **64**, 456–462 (2013)
4. Markovic, D.S., Zivkovic, D., Branovic, I., Popovic, R., Cvetkovic, D.: Smart power grid and cloud computing. Renew. Sustain. Energy Rev. **24**, 566–577 (2013)
5. Sliz-Szkliniarz, A.: Assessment of renewable energy-mix and land use trade-off at regional level: a case study for the Kujawsko-Pomorskie Voivodship. Land Use Policy **35**, 257–270 (2013)
6. Scleicher-Tappeser, R.: How renewables will change electricity market in next five years. Energy Policy **48**, 64–75 (2013)
7. Kiciński, J., Żywica, G.: Numerical analysis of defects in the rotor supporting structure. Adv. Vibr. Eng. **11**(4):297–304 (2012). ISNN 0972-5768
8. Kiciński, J., Żywica, G.: The numerical analysis of the steam microturbine rotor supported on foil bearings. Adv. Vibr. Eng. **11**(2):113–120 (2012). ISNN 0972-5768
9. Kiciński, J., Żywica, G.: Computational model and strength analysis of the steam microturbine with fluid-film bearings. In: Proceedings of the International Conference on Vibration Engineering and Technology of Machinery VETOMAC-VIII, Gdańsk, 3–6 Sept 2012, VM 2012-10005, pp. 333–342. ISBN 978-88237-61-4
10. Kiciński, J.: New method of state analysis and diagnostics of power micro-devices. Sci. Probl. Mach. Oper. Maintenance **1, 46**(165), 57–69 (2011) (National Research Institute, Random)

Chapter 2
Introduction to Microturbines Vibration Analysis

Abstract In this chapter general information is given concerning the dynamic properties of microturbines and the assumed methodology of research was discussed. In order to clarify the dynamic phenomena occurring in such machines, the theoretical principles of vibration analysis of mechanical facilities were integrated in this chapter. Afterwards, the rotating systems' vibration taking into account the type of rotor support and phenomena occurring in bearings that are capable of high-speed rotation were discussed in more depth. Different methods of analysis of rotating systems were characterized, enabling a theoretical description of real systems while guaranteeing acceptable level of accuracy. In the final part of the chapter, the examples of modern tools to support the engineering design process and experimental studies were presented. In regard to this, the capabilities of our own software allowing multiaspect analysis of fluid-flow machines and the capabilities of selected software available commercially were discussed. The test rig specially adapted for examinations in the field of rotor dynamics was also described. The test rig characteristics was supplemented with a description of the test equipment enabling dynamic tests of small-sized and high-speed rotors.

2.1 Characteristics of Microturbines

Since this monograph deals with modelling and experimental research of rotors and bearings used in energy microturbines, this subsection provides a concise description of such machines. This is important in view of the fact that devices of this type were not until now mass produced nor used in the national economy, and their construction is not widely known. Microturbines, which this book concerns, have a power up to a few 100 kW. Thanks to advanced electronics and automation, their maintenance-free start-up and operation at the established parameters is possible, with synchronization to the commercial power grid. Microturbines of such small power are most commonly built as gas turbines [1], and usually an assembly of a single-stage radial compressor and single-stage radial turbine placed on a common shaft constitutes their operational element (Fig. 2.1). Steam

J. Kiciński and G. Żywica, *Steam Microturbines in Distributed Cogeneration*,
DOI 10.1007/978-3-319-12018-8_2

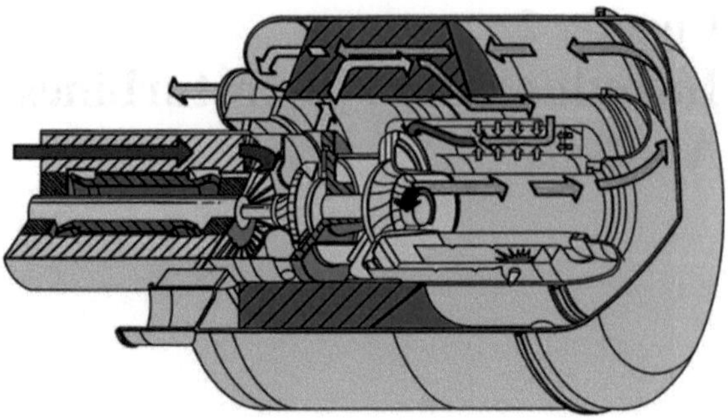

Fig. 2.1 Diagram of a gas microturbine with a power of 30 kW, from the American company Capston [4]

microturbines [2] of a power of several kW, which will be discussed in the following chapters, are among the first of this type of solution in our country and also represent great novelty on a global scale. The dimensions of the completed design solutions of microturbines used for distributed cogeneration, along with the required equipment, do not exceed the sizes of typical household appliances, such as refrigerators and washing machines. The main advantages of microturbines are [1–3]: high efficiency (in the cogeneration system even up to 90 %), low weight and compact design, as well as a simple mechanical construction (few moving parts—mostly just one rotating shaft). The most commonly encountered fuel in gas turbines on the market include: natural gas, biogas, diesel or kerosene.

Gas foil bearings [4] have been used for bearing of the rotor in the solution shown in Fig. 2.1 (see Sect. 3.3). This figure also shows the circulation of the working medium. The blue arrow marks the inflow of cold air into the compressor. Compressed air (yellow), after initial heating in the regenerative heat exchanger (orange), is directed to the combustion chamber. During the next stage, the ignition of the mixture follows (red) and the hot exhaust gases drive the rotor of the microturbine. The exhaust gases come out (yellow) additionally heating the working medium already before the combustion chamber (orange). This principle of operation described concerns gas microturbines frequently used in the industry [4]. The works carried out in IMP PAN, however, concern another type of heat turbines—steam microturbines [2, 3]. The way in which these turbines operate is different, as in this case, the heat energy of steam is used to set the rotor in motion. The fluid-flow of the working medium (obtained in a steam boiler or a heat exchanger called an evaporator) in this case is associated with a decrease in its enthalpy (potential energy), which results in the rotating movement of the turbine rotor. The mechanical energy of the rotor is then converted in a generator into electricity, and the used steam of the working medium is then condensed. The heat that accompanies these processes can be used to heat living areas or heat water in households. The following diagrams present a cross-section of a steam microturbine developed within the framework of the project POIG.01.01.02-00-016/08 (Fig. 2.2) and an ORC micro power plant with a microturbine (Fig. 2.3). This is a four-stage radial microturbine, which has two centripetal and two centrifugal stages. Two radial-thrust bearings are powered by steam of a low-boiling medium, and the generator placed between them is cooled by a water jacket.

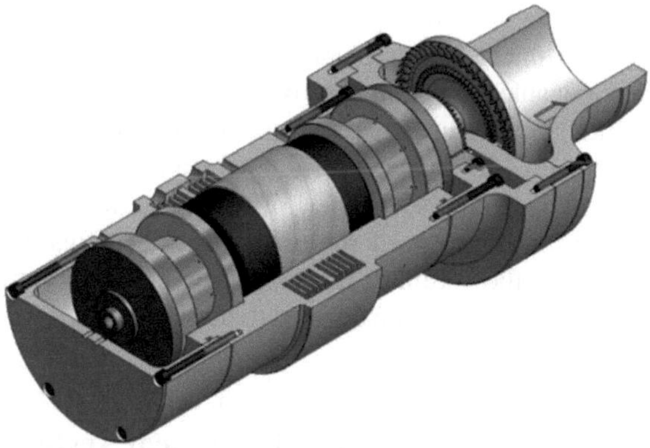

Fig. 2.2 Cross-section of a steam microturbine developed within the framework of the implemented research project [5]

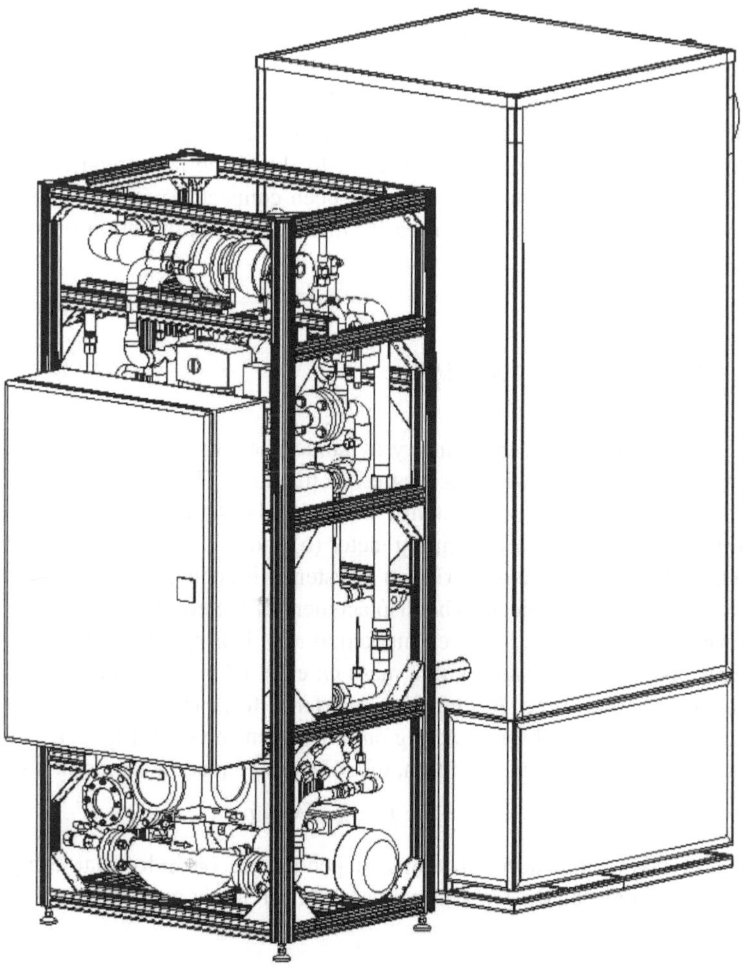

Fig. 2.3 ORC cogeneration micro power plant with a steam microturbine developed within the framework of the implemented research project [6]

The microturbine presented in Fig. 2.2 was designed and built with the use of an ORC cogeneration micro power plant in mind (Fig. 2.3). At optimal performance parameters it enables to obtain approx. 3 kW of electrical power.

The rotors of microturbines usually rotate at a rotating speed of over 20,000 rpm, with quite frequent pauses in operation. The temperature inside the microturbines typically amounts to a few 100 °C. In view of the above, there are many stringent requirements imposed on high-speed rotors and microbearings used in microturbines. Before launching, they are also the subject of advanced computer analyses. Among other things, this book will be devoted to these issues.

2.2 Introduction to Vibrations of Mechanical Systems

In available literature many books can be found which are dedicated to the vibrations of mechanical systems and dynamics of machinery [7–14]. These works contain the compendium of knowledge on the theoretical description and analysis of dynamic phenomena that occur in nature. For this reason, this chapter covers only selected, basic issues associated with the vibrations of mechanical systems. Therefore, this subsection only complements the introduction to further chapters and has been featured primarily in order to facilitate the interpretation of analyses results for those who have thus far not been connected with the dynamics of machines. An interested reader will easily find detailed information on the issues undertaken in the literature cited above.

2.2.1 The Equation of Motion

For the sake of simplicity, the primary dynamic phenomena will be discussed on the example of a system with one degree of freedom. This simple system represents the entire class of actual systems, which can be defined by four parameters: mass (m), elastic constant (k), damping factor (c) and exciting force ($F(t)$). In addition to this, the phenomena occurring in a system with one degree of freedom also apply to more complex objects. The displacement of the considered system, which can move in one direction, has been marked as $x(t)$. During vibrations, the following forces operate on a system with a mass m: external exciting force ($F(t)$), gravity force (G), inertial force (B), damping resistance force (S) and reaction (R). The scheme of the system and forces acting on it has been shown in Fig. 2.4.

Based on the equilibrium condition, the following equation can be written:

$$F + G + B - R - S = 0 \tag{2.1}$$

which after taking into account simple dependencies on individual components [8–14] can be presented as a second order linear nonhomogeneous differential equation:

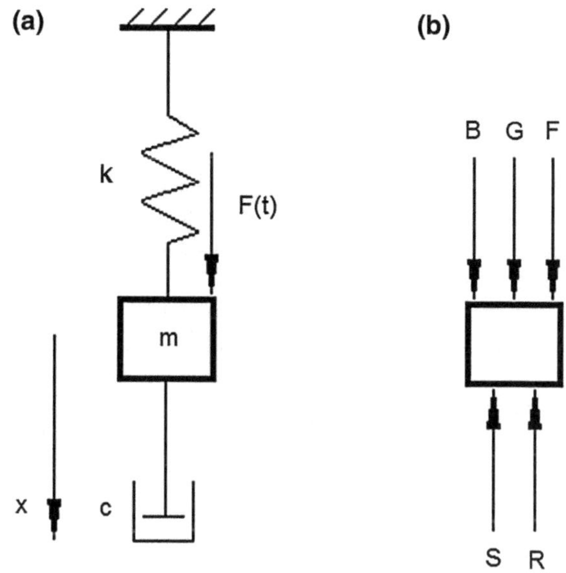

Fig. 2.4 System with one degree of freedom. **a** System scheme, **b** forces acting on the system

$$m\ddot{x} + c\ddot{x} + kx = F(t). \tag{2.2}$$

By dividing both sides of the equation by m and using appropriate formulas for free frequency (ω) and the unit damping factor (h) in the form of:

$$\omega = \sqrt{\frac{k}{m}},\ h = \frac{c}{2m},$$

the general differential equation of system vibrations with one degree of freedom is obtained:

$$\ddot{x} + 2h\ddot{x} + \omega^2 x = \frac{F(t)}{m}. \tag{2.3}$$

2.2.2 Free Vibrations

Where the external exciting force F(t) = 0, there are free vibrations of the system, which are described by a specific form of the Eq. (2.3):

$$\ddot{x} + 2h\ddot{x} + \omega^2 x = 0. \tag{2.4}$$

Further solution depends on the value of the damping. In the absence of damping in the system, the above equation is simplified to the form describing harmonic vibrations of a constant amplitude and period:

$$\ddot{x} + \omega^2 x = 0, \tag{2.5}$$

which after appropriate transformations [8] allows to determine the frequency and form of free vibrations in an non-damped system.

For systems in which damping cannot be ignored, the Eq. (2.4) is solved with a variety of methods depending on the value of damping. There are three variants of damping: supercritical ($h > \omega$), critical ($h = \omega$) and subcritical ($h < \omega$). A detailed introduction to each of these cases can be found in the literature cited earlier concerning the theory of vibrations.

If the damping in a system compared to spring force is large (supercritical or critical), damped vibrations occur. A deflected body then asymptotically seeks to the position of balance. Such movement is called by convention aperiodic damped vibrations [12] (although it does not have the nature of a vibrating movement). With a low damping (subcritical), the system is also characterized by vibrations of a continually decreasing amplitude, but the amplitude tends to zero when the time of vibrations grows to infinity.

2.2.3 Damping of Vibrations in Mechanical Systems

In order to evaluate the value of damping, a logarithmic decrement of damping is implemented. It expresses a natural logarithm of the ratio of absolute values of subsequent amplitudes of vibrations:

$$\delta = \ln \frac{|a_n|}{|a_{n+1}|}. \tag{2.6}$$

The logarithmic decrement of damping is a constant value, independent of time and amplitude.

The damping value is of particular importance with resonance, because only this feature of a system limits the amplitudes of vibrations. In mechanical systems, damping may be different in nature and come from a variety of sources. The most important practical significance is provided by: material damping, structural damping, which is accompanied by friction, and external damping (aerodynamic and hydrodynamic) of medium in which the system operates. Due to the fact that entire chapters of books and monographs [8, 15, 16] have been devoted to damping of vibrations, these issues shall not be discussed further here.

During numerical calculations and experimental studies, which are the subject of subsequent chapters, all of the causes of damping of vibrations mentioned above have been included. During analyses of hydrodynamic foil bearings, damping of lubricant film is very important. When analyzing foil bearings, in addition to aerodynamic or hydrodynamic damping of lubricant film, it is necessary to take into account material damping and friction between parts of the bearing. In order to properly describe the dynamic properties of any rotor, it is necessary to take into account the material damping in the shaft. Therefore, all of these phenomena are of great importance and cannot be omitted during the studies conducted.

2.2.4 *Forced Vibrations*

For practical reasons, a very important case of vibrating movement is vibrations forced by an external force, the value of which changes periodically according to the equation:

$$F(t) = F_0 \sin \omega t, \tag{2.7}$$

where: F_0 is the amplitude of the exciting force, ω is the angular frequency. The previously provided differential Eq. (2.3) in this case assumes the form:

$$\ddot{x} + 2h\ddot{x} + \omega^2 x = \frac{F_0}{m} \sin \omega t, \tag{2.8}$$

This is a nonhomogeneous differential equation, describing the forced vibrations of a system with one degree of freedom. The solution of the equation consists of solving a general homogeneous equation and a specific solution corresponding to the right side.

Any forced vibrating movement consists of the combination of forced vibrations and free vibrations of the system (depending on the initial conditions and the amplitude of force), while free vibrations have their own decreasing nature. As a result of the occurrence of damping, after some time, the forced component of the movement is of crucial importance. The nature of the transitional process depends on the ratio of exciting force frequency and the frequency of free vibrations. The interested reader will easily find details concerning the solution of the Eq. (2.3) in literature on the subject [8–16]. Noteworthy is the fact that the method of solving the equation above depends on the value of damping. An analysis of the solutions allows, among others, to draw the resonant curves of the system, which have practical significance.

2.3 Vibrations of Rotating Systems

This section will discuss the issues concerning the basic dynamic properties of rotating systems the type rotor—bearings—supporting structure. This subsection should be treated as an introductory part to the subject of dynamic analysis of microturbines. Because the issue concerning the dynamics of rotors is very extensive, it has been focused only on selected, very general matters. It will greatly facilitate the interpretation of results of the studies presented in the subsequent

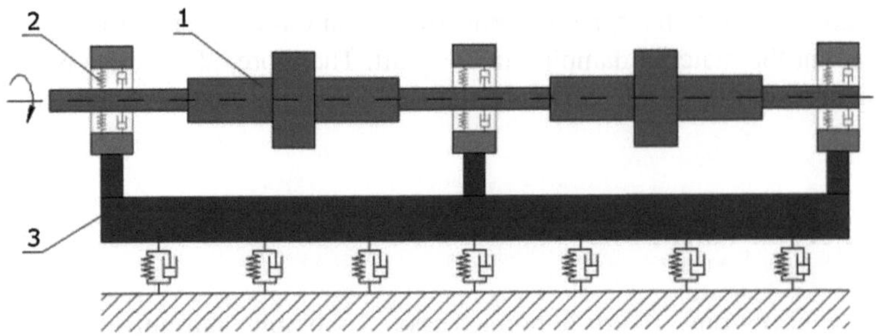

Fig. 2.5 Basic systems of a rotating machine. *1* Rotor, *2* bearings, *3* supporting structure

sections of the monograph for those who have not thus far dealt with this issue. It shall also help systematize specialized terms used repeatedly in the following chapters.

Each rotating machine consists of a number of complex, interconnected subassemblies. From the point of view of mechanics, three basic systems can be distinguished here (Fig. 2.5):

- line of rotors;
- bearings;
- supporting structure.

The line of rotors can consist of multiple sections of shafts interconnected e.g. by clutches. Within the concept of "line of rotors", also all the working elements set on particular sections of the shafts are taken into account (e.g. toothed wheels, rotor discs, generator). The term "bearing" refers to various types of rotor bearing systems together with the elements necessary for their functioning (e.g. seals). From the point of view of the dynamics of the system, in the case of slide bearings, it mostly refers to the flow-through load-bearing layer, while for rolling bearings these will be rolling elements working in conjunction with the rings. The supporting structure of rotating machines means the entire system on which the bearing line of rotors rests. This definition includes elements such as: bearing pedestals, frames, supports, bodies, vibration isolators and even foundations.

Each of these systems may constitute a separate research problem. It is very important that couplings occur between particular systems of rotating machines [17–25]. Thus dynamic properties of individual systems have an impact on the operational conditions of other systems. For example, changing characteristics of the supporting structure may have an impact not only on the work of the bearings, but also on the rotating shaft, and thus on the vibrations of the whole system. Due to the fact that a fluid-flow machine constitutes one coherent whole, its selected subsystems should not be analyzed separately from the others. It is necessary to use methods that will provide a consistent description.

A characteristic feature of the slide bearings commonly used in rotating machines is their non-linearity [26]. This means that to describe their functioning

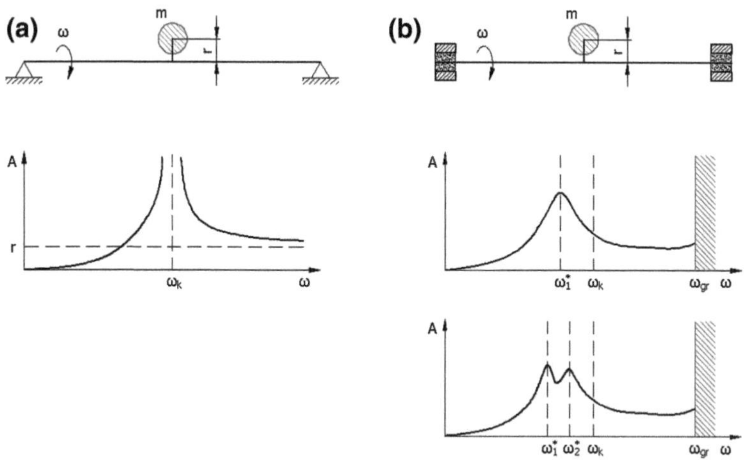

Fig. 2.6 The impact of slide bearings on the course of the vibrations amplitude of the rotor. **a** Rotor supported on rigid prisms, **b** rotor with slide bearings

in a wide range of rotational speeds, it is necessary to use non-linear models, which enable to generate non-elliptic trajectories of vibrations of system elements. The possibilities of linear models in which it is assumed that the displacements are proportional to the operating forces, in this case, work well only for stable operation of the bearing. To describe the properties of sliding bearings, so-called coefficients of stiffness and damping of lubricant film can be used. Characteristic of slide bearings is that the values of these coefficients depend not only on the instantaneous position of the journal within the oil clearance of the bearing, but also on the instantaneous speed of changes of this position [26]. It is also worth emphasizing how strong an impact these types of bearings have on the characteristics of a rotating shaft. Figure 2.6 demonstrates a comparison of the course of the relative vibrations amplitude journal-sleeve for a rotor supported on rigid prisms and fitted with slide bearings. In other words, the case with prisms can be treated as a system with rolling bearings. Characteristics correspond to the case where the source of forces is the imbalance of the rotor with a mass m on the radius r.

By analyzing these figures, it can be concluded that [17, 26]:

- slide bearings cause a displacement of the resonance speed in the direction of lower rotating speeds;
- the emergence of two resonant speeds is possible;
- thanks to the damping properties of the sliding bearings, the vibrations amplitude in the resonant range is of limited value;
- a border of stable functioning of the system occurs, above which the amplitude of vibrations can reach very high values (only limited by values of bearing clearance).

2.3.1 Hydrodynamic Instability

The aforementioned characteristics of rotors fitted with slide bearings determine that these systems require particularly detailed analysis. Such bearings can be the source of a very dangerous phenomenon, namely hydrodynamic instability. Beyond a certain rotating speed (also known as the hydrodynamic stability limit), we are dealing with the development of oil whirls and whip. As it has been mentioned earlier, to describe the properties of slide bearings, the coefficients of stiffness and damping of lubricant film can be used.

The anisotropy of the main stiffness coefficients c_{11} and c_{22} (horizontally and vertically) causes the emergence of two resonant speeds of the rotor. The source of the occurrence of hydrodynamic instability, thus oil whirls and whips, is the anisotropy of so-called cross-coupled stiffness coefficients c_{12} and c_{21}. These factors determine the ratio of increment of lubricant film reaction components (ΔW_x, ΔW_y) to components of displacement of the journal (Δx, Δy) caused by the increment of external force (ΔP). In the case of cross-coupled coefficients, the increment of lubricant film reaction in the direction perpendicular to the direction of displacements of the journal is taken into account. The inequality of cross-coupled stiffness coefficients $c_{12} \neq c_{21}$ is a characteristic feature of all slide bearings. This type of anisotropy does not occur in other typical mechanical systems. More information about the theoretical description of slide bearings can be found in literature (e.g. [17, 19, 20, 26]); here, however, we shall look closer at the phenomenon of hydrodynamic instability itself.

When discussing hydrodynamic instability, it should be pointed out that during experimental studies of the physics of this phenomenon, a number of various problems have been encountered; therefore, its nature to some extent can be recognized through the use of advanced computer analysis. This type of instability can manifest itself in two ways, through so-called small oil vibrations (also known as oil whirl) and large oil vibrations (oil whip). The shape of the trajectory of the centre of the journal and the quantity and distribution of characteristic points on the trajectory of vibrations are linked to subsequent phases of development of this phenomenon (so-called phase markers). These points indicate the position of the journal's centre at the beginning of the next two rotations of the rotor. Figure 2.7 presents subsequent phases of development of hydrodynamic instability, i.e.: the area of stable functioning, the area of small oil vibrations (oil whirl), the transition area and the area of large oil vibrations.

In terms of the stable functioning of the system, the shape of the trajectory of the journal's vibrations is usually similar to an ellipse. Beyond the border of stability, a single trajectory of vibrations begins to separate into two loops, and on the results of harmonic analysis a stripe begins to appear which corresponds to the so-called half-component. Then, within the transitional area, peak changes occur in the value of the vibrations amplitude and the separation of the trajectory takes place (it becomes an open curve). The next phase is already large oil vibration, at which the trajectories take the form of chaotic waveforms. The emergence of a

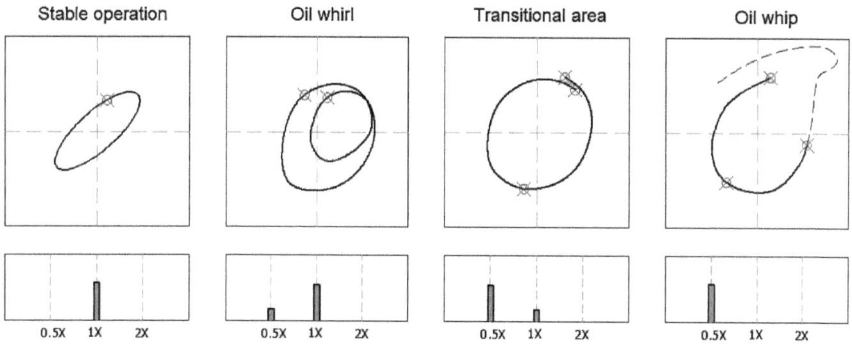

Fig. 2.7 Subsequent phases of development of hydrodynamic instability—changes in the trajectory of vibrations and the results of harmonic analysis

developed phase of hydrodynamic instability during the functioning of a machine in the form of large oil vibrations, in addition to the significant increase in the level of vibrations of the system's elements and disruption of working processes, can lead to serious failures, and even the total destruction of the machine.

From a practical point of view, it should be added that hydrodynamic instability is characteristic of lightly loaded bearings and it is revealed through increased vibrations of bearing journals at a specific rotating speed of a rotor. However, there are many other reasons, which can cause the occurrence of hydrodynamic instability. The most common are: increase of the viscosity of a lubricant, bearing displacement, change of properties of bearing supports, change of bearing clamp, and even imbalance and crack of the rotor [26].

The discussed rotating systems, and in particular systems with slide bearings, are very difficult to describe theoretically. Phenomena which are still the research subject of many scientific teams occur in them, and in the development of calculation models certain simplification assumptions are still used. Even more difficulties are encountered when developing models of whole machines with rotating elements. This is confirmed, for example, by failures that occur every so often, which happen even in the case of very carefully designed and analyzed constructions of rotating machines—so-called critical machines. For these reasons, among others, the theoretical models of fluid-flow machines are developed for the analysis of a particular group of issues, broken down e.g. by: strength, kinetostatic, dynamic or thermal phenomena. This will be explained further later in this chapter.

2.3.2 Kinetostatic Analysis

In kinetostatic analysis, variable dynamic forces acting on the rotating system are not taken into account. The calculations of the system take place assuming constant coefficients of stiffness and damping for a given rotational speed, taking into

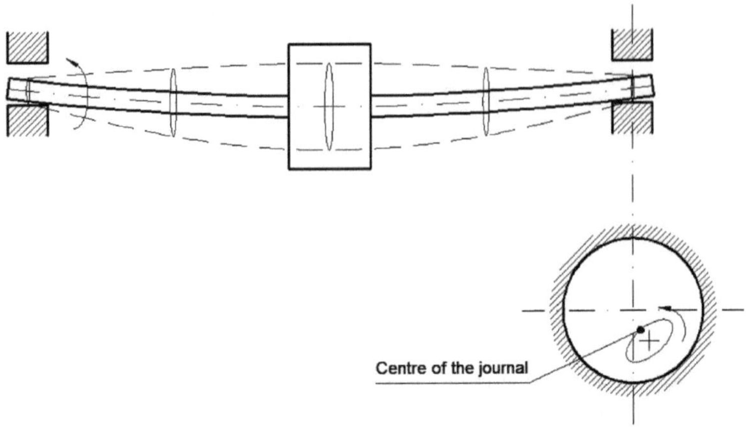

Fig. 2.8 The schematic representation of the phenomena occurring in the rotor-bearing system during kinetostatic analysis

account the rotation of the rotor without precession around the position of balance. In the case of complex multi-supported rotating machines, working in a wide range of rotational speeds, these are far reaching simplifying assumptions, enabling to determine solely initial characteristics of the system.

The main objective of this analysis is to determine the reactions at points of supporting the rotor, the kinetostatic deflection line and position of bearing journals, assuming that they do not vibrate around the position of balance. The simplifications adopted during the kinetostatic analysis also allow to determine certain characteristics concerning bearings, including: pressure distribution in the oil clearance, coefficients of stiffness and lubricant film damping, as well as the mentioned earlier position of the journal within the clearance gap. Kinetostatic analysis also enables the simulation of thermal processes occurring in the bearing node. During kinetostatic analysis, the force of gravity and other static forces acting on the considered system are taken into account, which are unchanging in time. The scheme of the rotating system including the kinetostatic phenomena has been shown in Fig. 2.8.

2.3.3 Analysis in the Scope of Linear Dynamics

During the dynamic analysis in the linear scope, periodic forces acting on the rotating system are taken into account, in particular, the harmonic strength caused by an unbalanced rotor. The objective of a dynamic analysis in the linear scope is to determine the trajectory of vibrations of established elements of the system, in particular the trajectory of vibrations of bearing journals. The operational parameters of the system are in this case determined using constant coefficients of

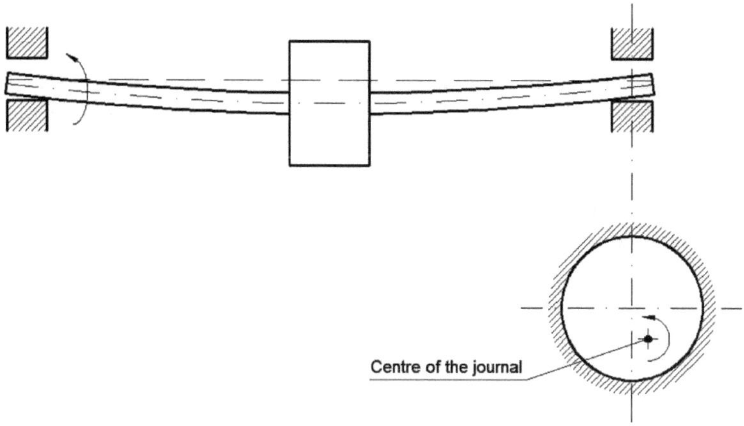

Fig. 2.9 The schematic representation of the phenomena occurring in the rotor-bearing system during dynamic analysis (linear vibrations)

stiffness and damping of bearings for a given speed, which enables to determine, at most, the elliptical trajectory of vibrations (Fig. 2.9).

As the result of calculations for slide bearings, the distribution of pressure and coefficients of stiffness and damping of lubricant film (main and diagonal) can be determined. As in the case of kinetostatic analysis, during calculations the modified shape of the oil clearance can be included, determined on the basis of the distribution of pressure (such a situation occurs e.g. in foil bearings). During dynamic analysis in the linear scope, for every rotating speed of the journal, dynamic coefficients of stiffness and lubricant film damping are determined (Fig. 2.10).

To define the dynamic properties of the bearing node, it is necessary to know the four coefficients of stiffness and damping of lubricant film, as well as two coefficients of stiffness and damping of supporting structure. These factors may be determined

Fig. 2.10 The scheme of the numerical model of the bearing taking into account the elasto-damping properties of lubricant film and supporting structure

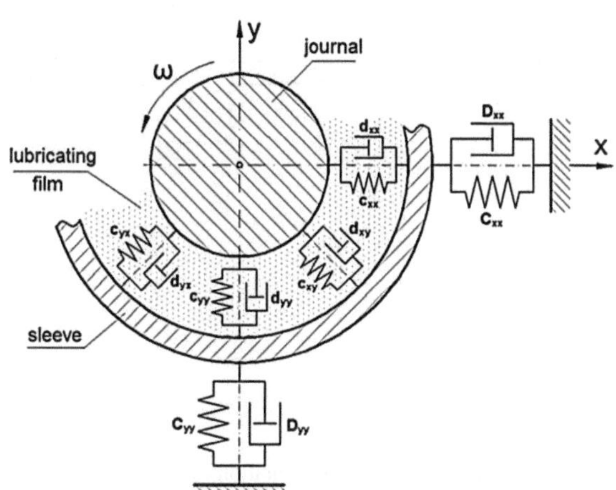

experimentally, as well as through numerical calculations. Thanks to this approach, the shape of the trajectory of vibrations of the system takes into account the characteristics as a whole of supporting the journal of the shaft in the bearing node.

A very important limitation of this method of dynamic analysis in the linear scope is its validity only in a stable scope of functioning of the system, when we are dealing with so-called small vibrations around the points of static balance. If this condition is not met, the description methods for so-called non-linear dynamics should be used.

2.3.4 Analysis in the Scope of Non-linear Dynamics

The analysis of linear vibrations discussed earlier does not enable the analysis of all the phenomena occurring in rotating machines with slide bearings. Above all, it is not able to describe the phenomena accompanying the unstable functioning of bearings, when self-induced vibrations of the journal occur, caused by non-stationary dynamic flow forces of the lubricant factor and when the damping abilities of bearings disappear. These phenomena are accompanied by large, non-elliptical trajectories of vibrations of the journal (Fig. 2.11), which are impossible to obtain using a model with constant coefficients of stiffness and damping. Such phenomena usually occur at high rotational speeds. By using a non-linear description, practically any external forces can be taken into account in calculations, as the non-linear analysis of vibrations of rotating systems essentially depends on time of the analysis of undefined vibrations.

Taking into account dynamic phenomena described above entails the necessity of applying a complex theoretical model of analyzing the system. This is

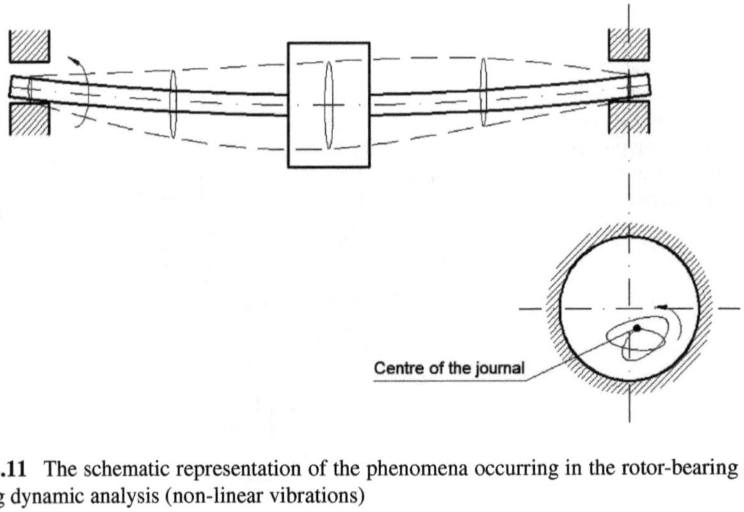

Fig. 2.11 The schematic representation of the phenomena occurring in the rotor-bearing system during dynamic analysis (non-linear vibrations)

particularly difficult in the case of such complex systems, like the rotor supported on slide bearings (lubricated with liquid or gas). Only an analysis that enables to generate complex trajectories of vibrations is able to show the advantages of certain types of bearings, such as all kinds of anti-vibration bearings and foil bearings.

Models enabling dynamic analysis in the non-linear scope are based on calculations carried out in the time domain. This is done by integrating the equations of motion in very small period intervals, enabling the precise drawing of trajectories of vibrations of the nodes in the analyzed system. During such an analysis, coefficients of stiffness and damping of bearings are determined in each analyzed moment of time, usually more than 100 times during each rotation of the rotor. As it has already been stressed, the development of such a model is very complex, both in theoretical and numerical terms. For this reason, dynamic analysis of rotating systems in non-linear scope is unavailable in commercial general CAE computer programs. Tools for non-linear dynamic analysis of rotors in recent years have become available as an additional option in some programs dedicated solely to calculating rotating systems [27]. It seems, however, that in order to fully benefit from the description and analysis of non-linear dynamic phenomena occurring in rotating systems, it is necessary to create and use the most flexible and open commercial computer programs.

2.4 Modern Research Methods and Tools

The workshop of the modern high-tech researcher of rotating systems should consist of both specialized computer software and modern measuring apparatus. This subsection will briefly discuss examples of research tools that were used during studies which are the subject of the following chapters.

2.4.1 Software

Currently, the market supplies many different types of computer software supporting engineering calculations, both of the general purpose and specialized—dedicated to solving specific technical problems. In the Dynamics and Diagnostics of Turbines Department (IMP PAN), there are several computer programs used for the computer analysis of rotary machines. The main calculation tool is the MESWIR environment created and developed for many years at IMP PAN [17]. In order to quickly solve engineering tasks from the scope of rotating machines, the commercial program MADYN 2000 is also used [27]. The problems from the field of solid mechanics, also including analyses of supporting structures and complex structures of foil bearings, are most frequently performed using ABAQUS [28]. One of the distinguishing characteristics of this program in terms of competition is the ability to conduct advanced analyses of non-linear systems, with regard to contact phenomena. Another generally available software used for structural and fluid-flow analyses is the package ANSYS Multiphysics [29]. The advantages of software included in this package can be used when conducting multidisciplinary analyses, e.g. coupled structural and fluid-flow analyses. The description of the functions and possibilities of the mentioned commercial software can easily be found e.g. on the Internet, therefore, here only the software package MESWIR developed at IMP PAN will be further discussed.

Programs from the series MESWIR (Finite Element Method in the analysis of the rotors) are used for the dynamic analysis of rotary machines in linear and non-linear scope. The MESWIR environment is an original achievement of IMP PAN in Gdańsk and constitutes the effect of many years of work of over a dozen people directly associated with the theory and practice of slide bearings and rotating machines [17, 30]. MESWIR environment has a modular construction, and its essential elements are:

- group of programs for the analysis of lateral and thrust slide bearings;
- group of programs to analyze the kinetostatic issues of the systems rotor-bearings-supporting structure;
- group of programs to analyze the issues of the dynamics of the systems rotor-bearings-supporting structure.

The structure of the package and names of individual programs have been shown in Fig. 2.12. Within each of the modules sets of tools were developed dedicated to

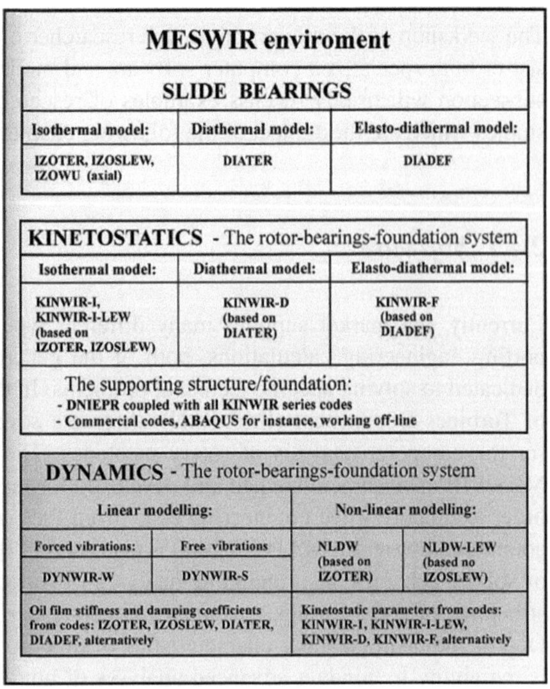

Fig. 2.12 The basic modules and names of software of the package MESWIR [17]

solve specific problems. Thanks to this, a detailed analysis of systems is possible in a very wide range of issues. Non-linear analysis plays a key role in the analysis of rotors with slide bearings, therefore, programs from the series NLDW are of particular importance here. Cooperation between particular applications ensures the compatibility of input and output data. Programs included in the package enable to perform an advanced analysis of various rotating machines, and especially large energy facilities, such as turbine sets. When taking into account the non-elliptical trajectories and multi-stripe spectres of vibrations of selected nodes of the system, it is possible to obtain symptoms corresponding to the defects modelled earlier, and then building a directory of diagnostic relations. In the case of critical machines, such understood diagnostics according to the model takes on particular importance, since very often it provides the only source of information about the change of operational characteristics in the case of the emergence of defects.

The starting point for a comprehensive analysis of the entire rotating machine is a set of bearing programs. This module is particularly relevant, since the slide bearings constitute a highly non-linear element of the entire system [17, 26]. The software developed within this module is based on three thermal models of a bearing, namely: isothermal, diathermal and elastodiathermal [26]. All the programs for analyzing bearings have the ability to record any geometry of the oil clearance, thanks to which it is possible to analyze both bearings with a circular-cylindrical clearance and anti-vibration (lemon type, offset, multi-wedge). It is also possible to take into account any non-parallelism of sleeve and journal axes of the bearing, as well as siphon pockets. All bearing programs calculate the coefficient of stiffness and damping of lubricant film using the perturbation method.

The issues of kinetostatics are a separate subject group in the MESWIR environment. As a result of the calculations, the line of kinetostatic bends of the rotor is obtained, as well as the reactions in particular bearing supports. For kinetostatic analysis, software of the series KINWIR has been developed, which set the basic parameters of a system based on the isothermal, diathermal or elastodiathermal model of the bearing. The following parameters are determined using them: the location of the centre of the journal in a plane of clearances, the hydrodynamic load carrying capacity of the bearing, friction force, oil demand, the minimum thickness of the oil film, the temperature of the oil liquid and coefficients of stiffness and damping of lubricant film in a point of static balance. The characteristics of the support structures, where the rotor is positioned, are introduced to the software series MESWIR in the form of a matrix of coefficients of stiffness and damping. The coefficients anticipated for bearing supports can describe both the dynamic properties of each of the supports and the couplings between them.

The parameters obtained on the basis of the discussed kinetostatic analysis are the starting point for calculating the dynamics of the system. It is possible to carry out an analysis in the linear scope (software from the series DYNWIR) and non-linear scope (software from the series NLDW). Software from the series NLDW makes possible the analysis of systems rotor-bearing-supporting structure at virtually any changes of the external load (physically justified). The added benefit of applying a non-linear description is the possibility to take into account all sorts of structural non-linearity, resulting from the presence of certain defects, such as a crack or misalignment of the line of rotors. The enormous potential of these programs enables them to be used as part of diagnostics according to the model for obtaining diagnostic symptoms [17]. The selected software from the series MESWIR has also been used in the development of the environment for the analysis of MESLOF foil bearings [31].

2.4.2 Test and Measurement Apparatus

Experimental studies of the dynamic properties of steam turbines discussed in this monograph were performed on test rigs where thermo-flow tests were carried out at the same time. Due to the fact that hermetic constructions of steam turbines powered by low-boiling mediums prevent the performance of certain measurements, they were performed on independent test rigs available in the Laboratory of Vibrodiagnostics IMP PAN. These additional tests concerned, above all, innovative bearing systems. The open and modular build of the universal test rig demonstrated in Fig. 2.13 and the test rig for foil bearings [33] enabled the performance of complementary dynamic and operational tests.

The main elements of a measuring system of test rigs are data acquisition units, which facilitate conducting measurements, signal conditioning and archiving results. On the global market today, there are at least a dozen vendors offering high-end computers and measuring cards, which enable, among others, to measure vibrations. In the Dynamics and Diagnostics of Turbines Department (IMP PAN), a multi-channel analyzer LMS Scadas Mobile is used to study the size of characteristic vibrations on actual objects and on test rigs (Fig. 2.14) [34]. This analyzer allows to carry out measurements in real time, as well as perform modal tests using up to 40 measurement channels. It is used for data registration using various types of measurement sensors, including: accelerometers, eddy current sensors, rotating speed sensors, strain gauges, thermocouples, etc. The manufacturer of the analyzer, the company LMS, is also a vendor of the software LMS Test.Lab [34], assisting the performance of measurements and interpretation of the results obtained.

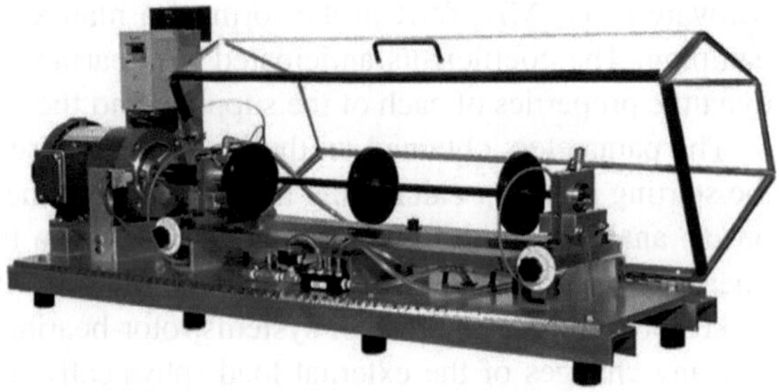

Fig. 2.13 Universal test rig for testing the dynamics of rotors and bearings [32]

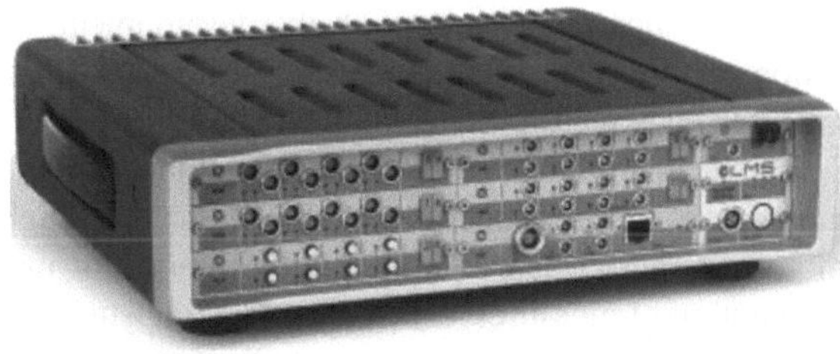

Fig. 2.14 Multichannel signal analyzer LMS Scadas Mobile [34]

Fig. 2.15 Miniature tri-axial accelerometer weighing 1 g

The full potential of data acquisition units can be used through modern sensors and transducers. Currently on sale there are many different types of vibration sensors available, as well as other physical parameters, which allows one to configure measuring apparatus even for very demanding and specific applications [35–37]. An example of modern vibration sensors can be miniature tri-axial accelerometers 356A03 from the company PCB Piezotronics (Fig. 2.15), the housing of which is made of titanium and which takes the form of a cuboid with sides of 6.4 mm, and the whole sensor weighs only 1 g [38]. These types of acceleration sensors can be used, for example, for modal tests of microrotors, where the mass of previously used tri-axial accelerometers caused large errors of measurement.

The necessary elements of a modern laboratory where research is conducted on the dynamic properties of rotors and other mechanical systems should be devices enabling precise forced vibration of the tested facility, the strength of a known amplitude and time course. Irreplaceable in this case are electrodynamic inductors of vibrations. Such inductors completed with an impact hammer make it possible to carry out modal tests even of very complex systems. The use of actuators of vibrations and an impact hammer to identify the dynamic properties of a construction has been discussed in Chap. 5.

In recent years, contactless measurements by laser has become more and more popular, especially in scientific application. This technique allows one to carry out vibration measurements using a contactless technique, using laser beams [39]. One of the advantages of this method is the elimination of distortions caused by the mass of sensors. The possibilities of laser vibrometers expand so-called optical derotators, which facilitate the measurement of vibrations on rotating elements [40]. Contactless measurements of vibrations also enable ultrafast digital cameras, recording images at a speed of more than 1 million frames per second. The software supplied together with such cameras enables a visual analysis of the vibrations of the tested systems. Both laser vibrometers and ultrafast cameras are currently very expensive apparatus. Along with the development of technology and the popularization of this type of equipment, its price should drop, which will provide wider access to such measuring techniques.

The availability of specialized test and measuring equipment causes that modern laboratories for dynamic and diagnostic tests of high-speed rotating machines are more and more often equipped with all sorts of instruments, which facilitate the preparation of measurements and conducting studies. It is becoming standard to use precision tests at the preparation stage of multiplane balancing machines of rotors or laser systems for alignment of shafts. In the case of studies on actual objects, compact, portable analyzers of the technical state of machines are also very useful (among others, enabling the spectral analysis of signals, the determination of characteristic frequencies, balancing in own bearings or prompt detection of damages to the bearings or cavitations). In detecting damages and improper operation of subassemblies of rotating machines, infrared cameras and sound analyzers are being used more and more often [41, 42]. The global, competitive economy and the continuous development of measuring techniques causes that the aforementioned investigator's apparatus increasingly constitutes the basic equipment of a laboratory.

References

1. Boyce, M.P.: Gas Turbine Engineering Handbook. Elsevier, Amsterdam (2006)
2. Bloch, H.P., Singh, M.P.: Steam Turbines. The McGraw-Hill Companies, New York (2009)
3. Soares, C.: Microturbines, Applications for Distributed Energy Systems. Elsevier, Amsterdam (2007)
4. http://www.wikipedia.org
5. Kozanecki, Z., Więckowski, S., Łagodziński, J., Marczyk, A., Borkowski, R., Kaczmarek, A.: Final assembly, balancing, start-up tests and dynamic tests of the second prototype of turbogenerator with aerostatic bearings using compressed air as the working medium. Int. Study IMP PAN, No. Arch. 59/11, Gdańsk, in Polish (2011)
6. Kiciński, J., Żywica, G., Niewiadomski, J., Cudakiewicz, M.: Design and technical documentation of the pre-prototype of micro CHP power plant with ORC. Int. Study IMP PAN, No. Arch. 467/2013, Gdańsk, in Polish (2013)
7. Dresig, H., Holzweissig, F.: Dynamics of Machinery. Springer, Berlin (2010)
8. Osiński, Z.: Theory of Vibration. PWN, Warsaw, in Polish (1978)
9. Gryboś, R.: Machine Vibration. Silesian University of Technology Publication, Gliwice, in Polish (2009)
10. Głowacki, H.: Engineering Mechanics, Dynamics. Oficyna Wydawnicza Politechniki Warszawskiej, Warsaw, in Polish (2001)
11. Piszczek, K., Walczak, J.: Vibrations in Machine Construction. PWN, Warsaw, in Polish (1972)
12. Łączkowski, R.: Vibrations of Components of Gas and Steam Turbines. WNT, Warsaw, in Polish (1974)
13. Wrotny, L.T.: Dynamics of Mechanical Systems. Oficyna Wydawnicza Politechniki Warszawskiej, Warsaw, in Polish (1995)
14. Gałkowski, Z., Kołacin, T., Kurnik, W., et al.: Mechanical Vibrations, Laboratory. Oficyna Wydawnicza Politechniki Warszawskiej, Warsaw, in Polish (1999)
15. Osiński, Z.: Mechanical Vibrations Damping. PWN, Warsaw, in Polish (1979)
16. Giergiel, J.: Mechanical Vibrations Damping. AGH Publication, Cracow, in Polish (1984)

17. Kiciński, J.: Dynamics of Rotors and Slide Bearings. IMP PAN Publication, Gdańsk, in Polish (2005)
18. Gosiewski, Z., Muszyńska, A.: Dynamics of Rotating Machinery. Wydawnictwo Wyższej Szkoły Inżynierskiej, Koszalin, in Polish (1992)
19. Rao, J.S.: Rotor Dynamics, 3rd edn. New Age International Publishers, New Delhi (2009)
20. Muszyńska, A.: Rotordynamics. Taylor & Francis Group, London (2005)
21. Gasch, R., Nordman, R., Pfutzner, H.: Rotordynamik. Springer, Berlin (2006)
22. Vance, J.M.: Rotordynamics of Turbomachinery. Wiley, New York (1988)
23. Żywica, G.: The analysis of the defects of the supporting structure in relation to the dynamic state of the turbomachine. Ph.D. thesis, The Szewalski Institute of Fluid-Flow Machinery PAN, Gdańsk, in Polish (2011)
24. Gasch, R.: Vibration of large turbo-rotors with fluid film bearings and an elastic foundation. J. Sound Vib. **47**, 53–73 (1976)
25. Walczyk, Z.: Dynamics of rotating power machines. Zeszyty Naukowe Politechniki Gdańskiej nr 482, Mechanika 64, Gdańsk, in Polish (1992)
26. Kiciński, J.: Theory and research of hydrodynamic radial slide bearings. Fluid-Flow Machines—Series, Tom 15. Ossolineum, Wrocław, in Polish (1994)
27. MADYN 2000 Documentation (2012)
28. ABAQUS Analysis User's Manual. Version 6.13, Dassault Systemes (2013)
29. http://www.ansys.com
30. Kiciński, J., Markiewicz-Kicińska, A.: The line of rotors and slide bearings. In: Kiciński, J. (ed.) Modelling and Diagnostics of Mechanical, Aerodynamic and Magnetic Impacts in Turbine Sets. IMP PAN Publication, Gdańsk, in Polish (2005)
31. Bogulicz, M., Żywica, G.: The MESLOF Software to Assist in the Analysis of Rotors Supported on Foil Bearings. Mechanik, 7/2013
32. Breńkacz, Ł., Bagiński, P., Żywica, G., Kiciński, J., Czoska, B.: Experimental studies on the microbearings and microrotors test rig. Determination of the reference characteristics for various configurations of the rotor-bearing system with slide and rolling bearings. Int. Study IMP PAN, No. Arch. 51/2013, Gdańsk, in Polish (2013)
33. Kiciński, J., Żywica, G., Bagiński, P., Niewiadomski, J.: Development of the method of experimental studies of gas foil bearings under thermal instability conditions. Development of the concept and technical documentation of the test rig. Int. Study IMP PAN, No. Arch. 987/2013, Gdańsk, in Polish (2013)
34. http://www.lmsintl.com
35. Barszcz, T: Machine Diagnostics and Monitoring Systems. Wydawnictwo Instytutu Technologii Eksploatacji, Cracow, in Polish (2006)
36. Barszcz, T., Urbanek, J.: Monitoring and Diagnostics of Turbomachines. Wydawnictwo Instytutu Technologii Eksploatacji, Cracow, in Polish (2008)
37. Kucharski, T.: Mechanical Vibration Measurement Systems. WNT, Warsaw, in Polish (2002)
38. http://www.pcb.com
39. Dwojak, J., Rzepiela, M., Struzik, I.: The use of a laser vibrometer for the operational diagnostics of electric motors based on own experience. Maszyny Elektryczne Nr 89/2011, in Polish
40. http://www.ects.pl
41. Kokociński, J.: Vibroacoustic Diagnostics of Machines. Energetyka Cieplna i Zawodowa, 11/2009, in Polish
42. Czech, J.: Selected Methods and Tools for Diagnostics of Technical Objects. Utrzymanie Ruchu, 1/2011, in Polish

Chapter 3
Unconventional Bearings

Abstract This chapter has been devoted to unconventional bearing systems for microturbine rotors. The introduction provides basic features and requirements imposed on bearings, which can be used in high-speed turbomachines. The chapter includes a proposal to divide the set of all bearings depending on their physical operating principle and on their inner structure. It also provides a comparison of selected properties of various bearing systems. Then, in the following sections of Chapter three, the methods of modelling and sample analysis results of nontypical bearings lubricated with low-boiling mediums and foil bearings were discussed. With respect to the foil bearing, a step-by-step process of implementing complex geometry and the FE model was described. Modelling of the fluid-structure interaction matter has received much attention. It also presents the results of experimental verification of the numerical models and examples of their use. At the end of the chapter basic information about the other, non-conventional bearing systems, including: high-speed rolling bearings and magnetic bearings was given.

Modern, high-performance fluid-flow machines very often operate in difficult conditions. For example, the rotors of microturbines usually rotate at a rotating speed of well over 10,000 rpm, with quite frequent pauses in operation. The temperature inside the microturbines typically amounts to a few hundred degrees Celsius. In view of the above, there are many stringent requirements imposed on microbearings used in rotors of microturbines. This forces constructors of such machines to apply innovative methods of bearing shafts, which on the one hand will ensure the proper functioning of the machine, and on the other, will be acceptable in technical and financial terms.

This chapter discusses various unconventional methods of fitting bearings in steam microturbine shafts, which can also be used in other machines. Because the dimensions of these types of bearings are very small, they are quite commonly called microbearings. Among the proposed solutions are low-boiling liquid-lubricated bearings, foil bearings and ceramic rolling bearings. At the beginning of the chapter, there is a subsection which is an introduction to the subject matter. Then, in respect to all solutions, their advantages and disadvantages were discussed, and practical

J. Kiciński and G. Żywica, *Steam Microturbines in Distributed Cogeneration*,
DOI 10.1007/978-3-319-12018-8_3

considerations associated with their use have been provided. For these types of bearings numerical models developed and used at IMP PAN have been presented, as well as examples of an analysis of rotating systems with such bearings. Getting familiar with this chapter should allow the reader to become acquainted with very broad possibilities in the scope of new solutions available in the field of bearing engineering.

3.1 General Information about Microbearings

Small, high-speed bearings have in recent times been a popular subject of research work in many institutes, both industrial and academic worldwide. Their popularity has grown along with the progressing miniaturization of constructions and development of so-called eco-energy. They are, in fact, one of the most important elements of microturbines—the basic machines used in distributed generation.

This chapter concerns bearings, which could be used in microturbines for the simultaneous production of electricity and heat. Of course, other applications of high-speed bearings are also known, which will not be discussed further here, such as [1]:

- gas turbine engines (air and traction drives);
- auxiliary equipment in nuclear energy (pumps, compressors, blowers);
- high-speed processing machines (drilling, grinding spindles);
- electric motors and generators of special purpose (in an atmosphere other than air, with the requirement of low levels of vibrations and noise);
- control and measurement equipment, gyroscopes;
- medical tools;
- machines for the textile industry, pharmaceutical and food industry.

Below (Fig. 3.1) a proposed classification of microbearings has been presented [2, 3]. This classification includes all commonly used types of bearings, while it has been focused on transverse bearings. Each of the groups of bearings can be further divided depending on, for example, construction materials or the lubricant medium used. And so, within rolling bearings, the following types can be distinguished: steel, hybrid (steel-ceramic) and ceramic. Divisions according to other criteria are also possible.

Currently on the market there are many solutions available, which are dedicated to high-speed bearing rotors of small nominal diameters. Therefore, it is necessary to adopt criteria of selecting appropriate bearings depending on the specific application, which will allow for the correct and long-lasting operation of the machine. A choice made in such a way should always be treated as preliminary, and to facilitate the elimination of certain constructions, which will enable to narrow down the area of acceptable solutions.

The most important criteria of selecting bearings, depending on their purpose, have been presented below. They are:

- acceptable rotational speed;
- load carrying capacity;

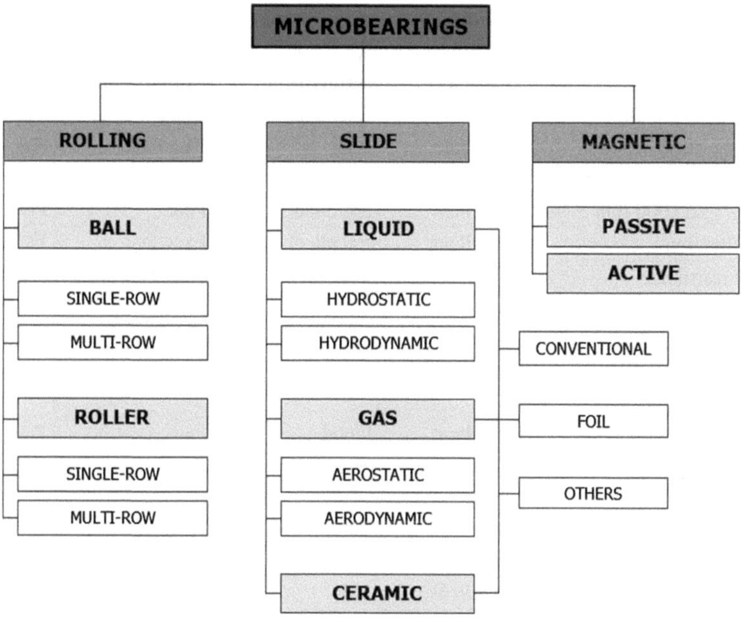

Fig. 3.1 The proposed division of microbearings used in microturbines [3]

- acceptable operating temperature;
- durability and reliability, ease of repair;
- precision and stiffness of bearings;
- vibration damping capacity;
- noiseless operation;
- resistance to external factors;
- availability;
- price.

Of course, the list of requirements set for microbearings is much longer and very often depends on the specifics of the considered case. Table 3.1 provides a summary showing the usefulness of different types of bearings (gas, oil, and rolling) depending on the requirements set before them. As it can be seen from the above table, most of the criteria are perfectly met by aerostatic gas bearings. These bearings have quite an important flaw, because the need to supply them with gas from the outside causes significant complication of the entire machine's construction. Worst in the set are rolling bearings, which principal defect in relation to microturbines is their quite low acceptable rotational speed.

The following points of this chapter discuss in more detail particular types of bearings. Much attention has also been paid to discussing the methods of their modelling and analysis.

Table 3.1 Comparison of selected characteristics of bearings [43]

Parameter		Air bearing aerostatic	Air bearing aerodynamic	Oil-bearing hydrodynamic	Ball bearing ang. contact
Accuracy of rotation		Excellent	Excellent	Good	Fair
Speed	<1,000 rpm	Excellent	Poor	Excellent	Fair
	<1,000–60,000 rpm	Excellent	Excellent	Fair	Good
	>60,000 rpm	Excellent	Excellent	Poor	Poor
Low vibration		Excellent	Excellent	Excellent	Fair
Shock resistance		Good	Fair	Excellent	Good
Frequent stop/starts		Excellent	Good	Excellent	Good
Low starting torque		Excellent	Fair	Good	Good
Long lifetime (>20,000 hours)		Excellent	Good	Excellent	Poor
Wide temperature range		Excellent	Good	Fair	Fair
Contamination to surroundings		Good	Excellent	Poor	Poor
Resistance to dust ingress		Excellent	Fair	Good	Good
High axial/radial loads		Good	Fair	Excellent	Excellent
High axial/radial stiffness		Excellent	Fair	Excellent	Good
Small space envelope		Fair	Good	Good	Excellent
Low heat generation		Good	Good	Poor	Good
Run in partial vacuum		Fair	Poor	Fair	Fair
Low running costs		Fair	Excellent	Good	Good

3.2 Bearings Lubricated with a Low-Boiling Medium

In slide bearings the cooperating surfaces of the journal and sleeve are separated from each other by a lubricant layer, thus enabling the rotation of the rotor. Depending on the construction, purpose and current working conditions of the bearings, dry, liquid or mixed friction can occur. In practice, mixed friction is most commonly encountered (low-speed rotating machines), as well as liquid friction. Almost always, however, it is sought to obtain liquid friction in the slide bearings. A lubricant medium of slide bearings can be liquids and gases, except that gases (e.g. air) are applied in the event of a small carrying capacity and high rotational speed (above 10,000 rpm). A wide variety of construction can be defined by the term "slide bearings", starting with classic hydrodynamic bearings and hydrostatic bearings, through gas bearings and to the innovative solution of foil bearings. This group includes transverse, longitudinal and transverse-longitudinal bearings. This part of the book focuses on hydrodynamic and hydrostatic bearings, thus those greased by a liquid medium.

Slide bearings are commonly used in the case of:

- large loads (even up to several MN);
- the presence of shock loads and need of vibration damping;
- high rotational speeds;

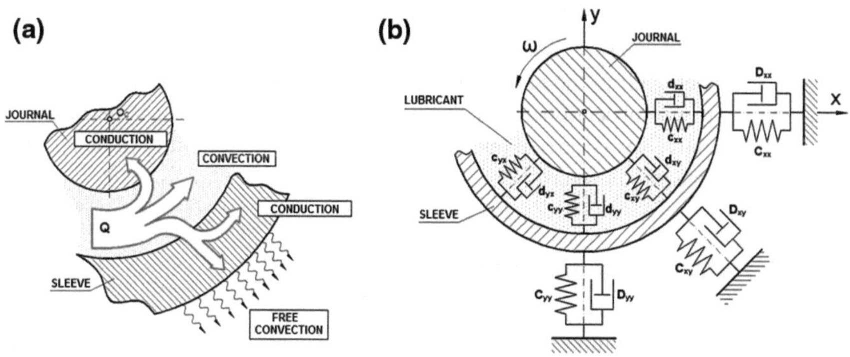

Fig. 3.2 Theoretical models of slide bearings. **a** Thermal model. **b** Dynamic model

- the need to apply split bearings;
- required noiseless operation.

It should be noted that slide bearings constitute a system that is extremely difficult to model in theory. Regardless of the type of construction and applied lubricant medium, a number of heat-flow processes occur in them, which affect the dynamic properties of the rotor (Fig. 3.2).

The conditions of friction present in slide bearings depend not only on the lubricant medium applied, but also on bearing clearances, materials used, the condition of cooperating surfaces, as well as external forces acting on the system. During operation of the bearing node large amounts of heat are let out (this does not apply to gas bearings), which has an adverse effect on the operation of the bearing, accelerating its wear. In order to reduce energy losses and wear of bearings, as it has been previously mentioned, in slide bearings it is most often sought to obtain liquid friction. It requires the fulfilment of several conditions, the most important of which are: a sufficiently high rotational speed, large affluence of oil of an appropriate viscosity, low roughness of cooperating surfaces and properly selected bearing clearance. In order to improve the working conditions of the bearings, powering the oil clearance with a lubricant medium under pressure is also applied. Both of these solutions, i.e. hydrodynamic and hydrostatic bearings will be discussed in the following subsections.

The proper operation of slide bearings depends largely on the properties of the materials from which the journal and sleeve are made. In order to obtain the most favourable operational conditions of the bearing, the material properties of friction steam should be selected properly. The main criteria in the selection of materials for slide bearings are: resistance to seizing, corrosion resistance, low coefficient of friction, mechanical and deformation strength, appropriate thermal properties, as well as good technological parameters, such as workability and ease of coating. Because slide bearings are used in a very wide range of rotational speeds and temperatures, the selection of appropriate materials is an individual matter that depends on the specific application of bearings.

An overview of the different types of bearings, also including slide bearings, has been provided in earlier works carried out within the framework of the scientific network Eko-Energia [2–5]. These works discuss in detail all sorts of innovative design solutions of bearing nodes and other methods of testing them. Much attention has also been paid to the review of materials used for the manufacture of bearings. The next two subsections shall discuss structural solutions of slide bearings, for which an analysis was carried out in order to assess the possibility of using them when building steam microturbines.

3.2.1 Hydrodynamic Bearings

Due to the fact that commonly used slide bearings are characterized by features that prevent their use at rotational speeds coming up to 100,000 rpm, in the case of rotor bearings in a microturbogenerator, it would be necessary to apply a special solution, in order to obtain such speeds. The basic problem with using slide bearings in high-speed rotors is connected with the appearance of the stability threshold of the rotor-bearings system. In the case of traditional slide bearings lubricated with oil, this limit is usually several thousand rotations per minute. Above that speed there are vibrations and oil whip, which can even lead to destroying the machine. At the Institute of Fluid-Flow Machinery of PAN in Gdańsk, a concept was created of using a low-boiling medium as the lubricant agent in slide bearings—the same as in the thermodynamic cycle of ORC micro power plant. Thanks to the use of such a medium, which is characterized by a very low viscosity in relation to traditional liquid lubricants, it is possible to obtain much higher rotational speeds, in the range of which microturbine usually operated. In order to administer the low-boiling medium factor into the lubricant clearing of bearings in liquid form, it would be necessary to plan exactly the location of collecting the medium from the thermodynamic circulation of the micro power plant (e.g. directly behind the condenser). The remainder of this subsection will present the structural solution of slide bearings lubricated by a low-boiling medium developed at IMP PAN. For such bearings, simulation calculations have also been carried out, which allow a preliminary assessment of their usefulness in the bearings of the rotor for the developed microturbines.

The basic dimensions of the proposed structural solution of slide bearings have been provided in Table 3.2. These are bearings with a circle-cylindrical clearance, for which the width-diameter ratio is 1. The viscosity of the lubricant medium in relation to bearing oil was approx. 1,000 times smaller. This allowed to carry out simulation tests of the rotor-bearings system (for the system shown in Fig. 3.3) in terms of speeds up to 100,000 rpm. During the tests programs from the series MESWIR were used [6].

Table 3.2 The basic dimensions of the proposed hydrodynamic slide bearings lubricated by a liquid low-boiling medium

Parameter	Unit	Bearing 1	Bearing 2
Nominal diameter	mm	10	10
Bearing width	mm	10	10
Radial clearance	mm	7.5×10^{-3}	7.5×10^{-3}

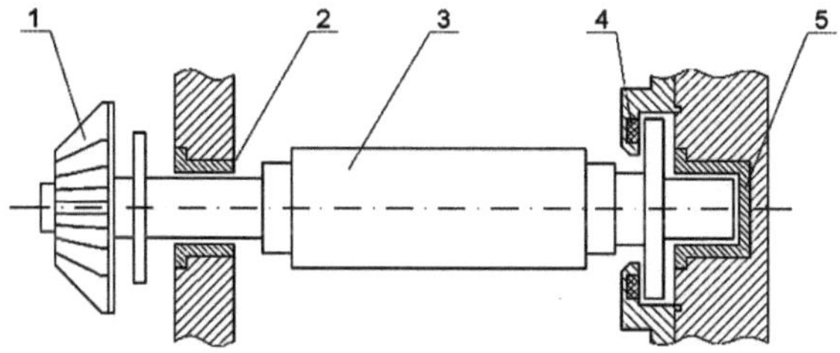

Fig. 3.3 Example of bearings on the turbogenerator rotor using slide and magnetic bearings. *1* blade disc, *2* lateral slide bearing, *3* generator rotor, *4* passive thrust magnetic bearing, *5* lateral-thrust slide bearing

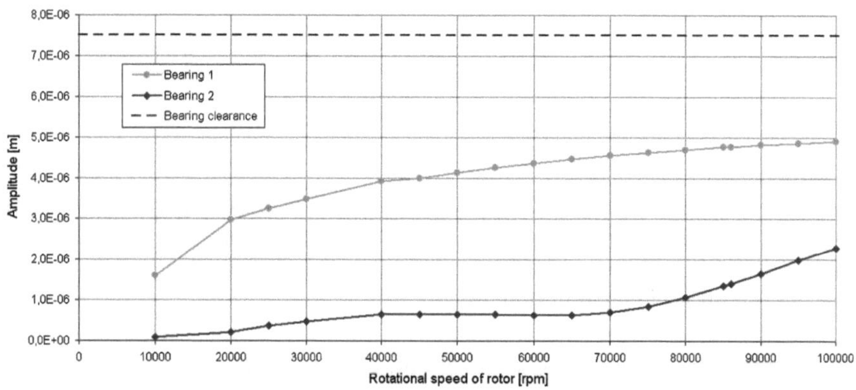

Fig. 3.4 The course of amplitudes of relative vibrations of journal-sleeve for the microturbine rotor with hydrodynamic slide bearings lubricated by a liquid low-boiling medium

The simulation tests performed provided results in the form of: course of amplitudes of relative vibrations of journal-sleeve, absolute vibrations of sleeve, distributions of pressure in the lubricant clearance and trajectories of the vibration of elements of the rotor-bearings system. It should be emphasized that due to the lack of information on the final solution of the fluid-flow part (which makes it impossible to determine the longitudinal forces on the shaft), only transverse forces were taken into account in the calculations. Chosen results of the studies have been presented in Figs. 3.4 and 3.5.

The conducted simulation tests confirmed the possibility of applying to the bearings of the turbogenerator rotor hydrodynamic slide bearings lubricated by a liquid low-boiling medium. As it is shown by the course of relative vibrations of bearing journals presented in Fig. 3.4, throughout the entire analyzed range of rotational speeds (from 10,000 to 100,000 rpm) the system worked steadily. This is confirmed also by the analysis of trajectories of vibrations for the selected rotational speed presented in Fig. 3.5. Similar trajectories of vibrations were obtained

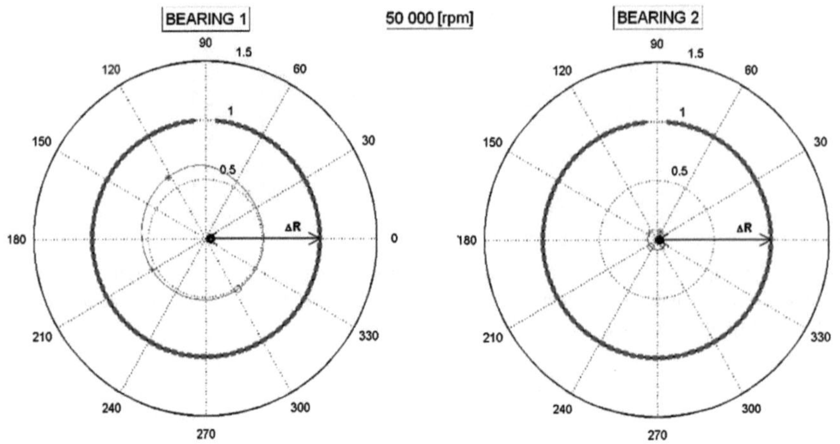

Fig. 3.5 The trajectories of the vibrations of journals of hydrodynamic slide bearings at a speed of 50,000 rpm

for other rotational speeds. Throughout the entire analyzed range, no signs of hydrodynamic instability were observed. Similar simulation tests were also performed for other widths of bearings, different values of bearing clearances and varying conditions of bearing supports [7].

During the practical implementation of the actual rotor-bearings system, according to the presented concept, the biggest problems associated with bearings should be expected in the range of ensuring ample lubrication of bearings by a low-boiling medium in the appropriate form. The low temperature of boiling of such mediums and the presence of a fairly high temperature in the fluid-flow part of the microturbine can be the cause of serious problems when building and operating an actual microturbine.

The next section of this chapter shall present the concept of slide bearings, in which, in order to improve the dynamic properties of the rotor-bearings system, additional siphon pockets and hydrostatic power were introduced.

3.2.2 Hybrid Bearings

Due to the fact that in the case of large steam turbines, introducing additional siphon pockets to the construction of slide bearings in many cases significantly improves the dynamic properties of the system and facilitates start-up, it is worth considering such a solution also in relation to microturbines. In hybrid bearings, the hydrodynamic lubricant wedge formed as a result of the relative motion of cooperating surfaces of the journal and sleeve, as well as the narrowing of the oil clearance, is additionally powered hydrostatically, by a lubricant medium under

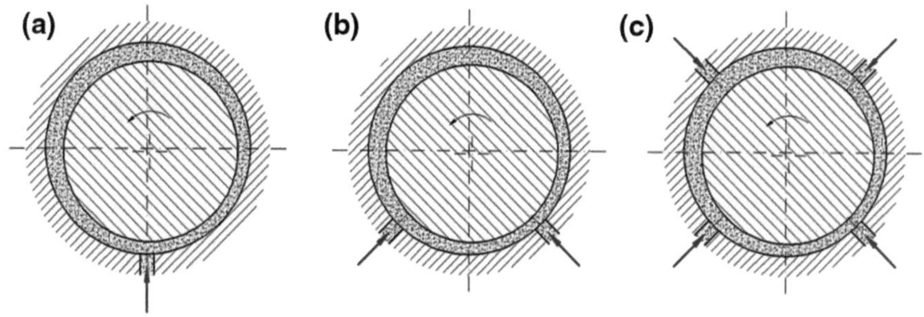

Fig. 3.6 Different variants of distributing siphon pockets

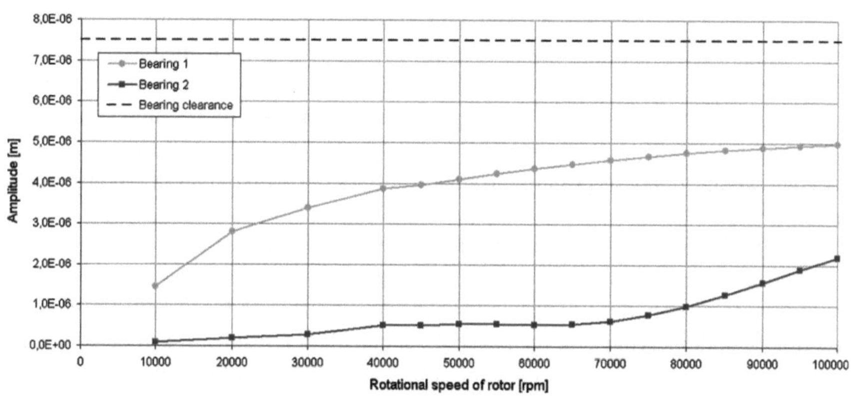

Fig. 3.7 The course of the amplitude of relative vibrations of journal-sleeve for slide bearings with 2 siphon pockets

pressure through the siphon pockets. In these types of bearings, the shape and distribution of siphon pockets are of great importance. At IMP PAN simulation tests were performed of the dynamic properties of the turbogenerator rotor set on hybrid slide bearings for several variants of arranging siphon pockets. The analyzed variants of distributing siphon pockets have been shown in Fig. 3.6.

Simulation tests, as in the case of hydrodynamic slide bearings, were carried out using programs from the series MESWIR [6]. The system presented in Fig. 3.3 was analyzed, and like in the previous case, the longitudinal bearings were not taken into account. The basic dimensions were compatible with Table 3.2, and the only difference in relation to the previously discussed bearings was the introduction of siphon pockets.

According to the obtained results of calculations (Figs. 3.7 and 3.8), it should be noted that, in the case of the analyzed turbogenerator rotor, as a result of introducing siphon pockets to the bearings, no substantial improvement of the system's dynamic properties were obtained. Despite the performance of comprehensive

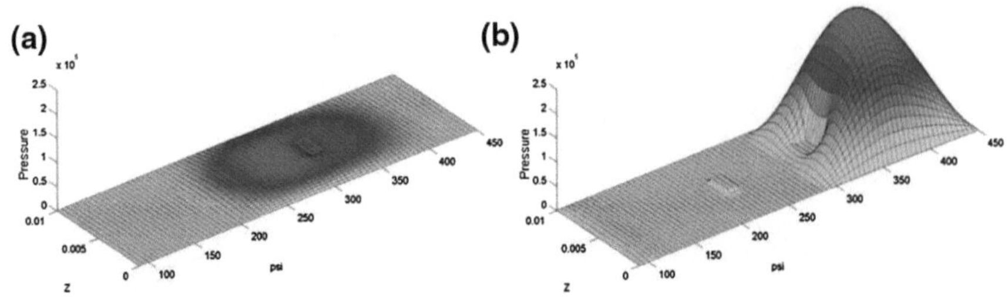

Fig. 3.8 The pressure distribution in the clearance gap of the hybrid bearing with four siphon pockets at a speed of 40,000 (**a**) and 100,000 rpm (**b**)

studies, in which an attempt was made at the selection of the optimal distribution of siphon pockets, their dimensions and values of power pressure, a substantial improvement in the dynamic stability of the system was not obtained [8]. Moreover, when selecting the optimal value of the power pressure, at higher pressure values, clear deterioration was noted of the operational conditions of bearings. Only relatively low values of pressure (several times lower than the maximum value in the lubricant wedge) made it possible to obtain results not worse than in the absence of siphon pockets.

The simulation analyses carried out showed that in the case of bearings of the microturbine rotor, the introduction of siphon pockets does not have a positive effect on its dynamic properties. Because such pockets significantly complicate the construction of slide bearings, on the basis of the calculation results, it can be concluded that there is no justification for their introduction in the case of the analyzed system.

Despite the fact that hydrodynamic and hydrostatic slide bearings are commonly used in the range of fairly low rotational speeds, it turned out that after introducing certain modifications, they can be used even at speeds ranging up to 100,000 rpm. This is possible thanks to the application of lubricant liquid of a low-boiling medium, which is characterized by a very low viscosity. This solution, however, involves the emergence of a number of practical problems, such as: providing relevant parameters of the lubricant medium supplied to the lubricant clearing of bearings, the need to use a special system solution powering the bearings, or the problem with the high penetration of low-boiling mediums by seals. Despite these drawbacks, these types of bearings seem to be an interesting alternative to other bearings used at high rotational speeds.

3.3 Foil Bearings

Scientific publications and information available on the Internet suggest that foil bearings are increasingly used in high-speed machines, especially those working at high temperatures [2, 9–11]. Bearings of this type belong to the group of slide bearings, however, in addition to the fluid-flow carrying layer, there is a structural

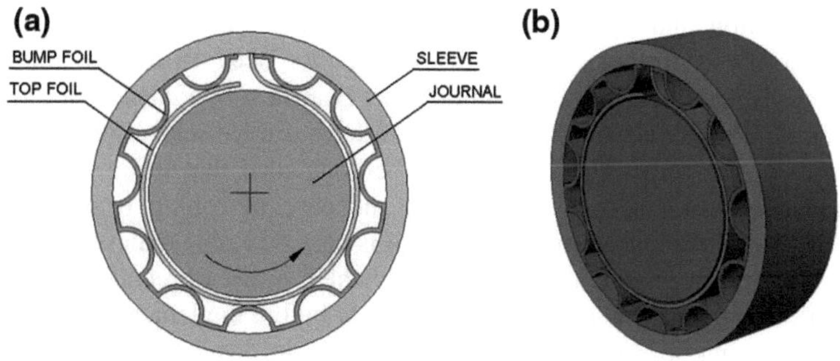

Fig. 3.9 Foil bearing structure. **a** Side view. **b** Isometric view

carrying layer in them, composed of an assembly of foils of considerable sensitivity. Thin, appropriately profiled foils, have very good damping properties. The foil, on which the journal rests, is called top or slide foil, and the foil found between the top foil and sleeve is called bump foil or supporting foil. Thanks to the use of such a solution, rotors supported in foil bearings are characterized by a much larger margin of stability. This is due to the fact that the shape of the oil clearance during its operation constantly adapts to the current conditions, and the poor damping properties of the lubricant film are compensated by the foil damping. The excess energy of the system is dispersed as a result of the deformations and friction between the cooperating elements of the foil assembly. The principle of building a typical foil bearing of the 1st generation has been shown in Fig. 3.9. In actual bearings, the foils are very thin (around 0.1 mm), and the radial clearance between the journal and the sleeve does not exceed a few thicknesses of the foil.

A few generations of foil bearings are distinguished. They differ in the way elasto-damping elements are arranged in the axial and peripheral direction [2]. In the latest structural solutions of foil bearings of the 4th generation, the top foil is covered by a coat of ceramic composite. Thanks to this, foil bearings can operate in extreme conditions (temperature approx. 800 °C, speed approx. 700,000 rpm). The successive generations of foil bearings were characterized also by an increasing carrying capacity. Bearings of this type work mostly as self-maintenance gas bearings, and air is usually used as the lubricant medium. In the case of gas turbines, which are a typical example of the application of foil bearings, it greatly simplifies the construction of bearing nodes, as additional powering by a lubricant medium is not necessary. Because properly designed foil bearings do not wear at a nominal speed, it is not necessary to replace them frequently, as it is the case with rolling bearings operating at a high temperature.

Currently, none of the world's manufacturers of foil bearings offers a ready catalogue of products. This is due to their specific properties. In this case, every implementation must be treated individually. A specific solution of foil bearings should be applied only in the machine for which they were designed. At the moment, any attempt to purchase foil bearings is connected with providing for their manufacturer, in addition to the basic dimensions, accurate data on the construction and operational conditions of the machine. The purchase of these bearings is irrevocably associated with the need to design, implement, and carefully test them by the manufacturer. Even a subtle change in the shape, thickness and material of the bump foil can have a huge impact on the dynamic properties of the rotor. Therefore, it can be concluded that we are dealing with a completely different approach than in the case of rolling bearings and oil lubricated slide bearings. Also due to this, it is very desirable to develop and apply reliable numerical models of foil bearings, which allow to reduce the costs and time-consuming experimental studies, necessary before implementation.

The following sections of this chapter discuss in detail the various stages of developing a foil bearing model, which was developed at IMP PAN. For a 3D and 2D model of the structural bearing layer, the most important information has been provided, concerning: creating the geometry, parameterization and developing a discrete model (FEM). Then, the initial results of analyses carried out using the developed models have been provided. In the development of numerical models of the structural bearing layer of a foil bearing, general purpose commercial computer programs were used that are available on the market. For the development of the parametric geometric model, the program Autodesk Inventor was used. While to develop the discrete model, the program ABAQUS was used. The cooperation of these programs took place thanks to the possibility of generating and loading files of the same extensions. Autodesk Inventor is a popular CAD program, allowing to create solid models and complex surface structures, as well as share files with most known programs for computer assisted design. ABAQUS is a highly respected FEM program, used in many academic and industrial centres. Its distinctive feature is the ability to analyze very complex systems, taking into account nonlinearity and contact phenomena. The model developed using the current versions of these programs will therefore use the latest computer technology, both in terms of modelling and engineering analysis.

3.3.1 Geometry of Foil Bearings

This subsection shall discuss the issues relating to the development of the geometric model of the foil bearing structure. So that the assembly of foils fulfils its role, it should have properly designed and implemented geometry. This applies particularly to the bumps of the supporting foil, which on the one hand provide adequate damping properties, and on the other, they cannot allow too large displacements of the journal.

The process of creating the geometry shall be discussed on the example of using the generally available program CAD. The primary requirement was, in this case, the parametrization of the model, enabling quick and easy change of geometry, depending on the current needs. The developed parametric model had to be adapted to carry out analyses of structural deformations of the carrying layer of the bearing, caused by static and dynamic loads.

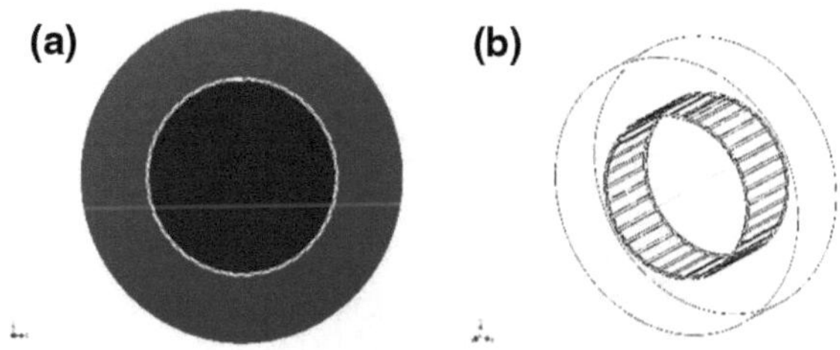

Fig. 3.10 The geometry of a 3D model of a foil bearing—assembly diagram

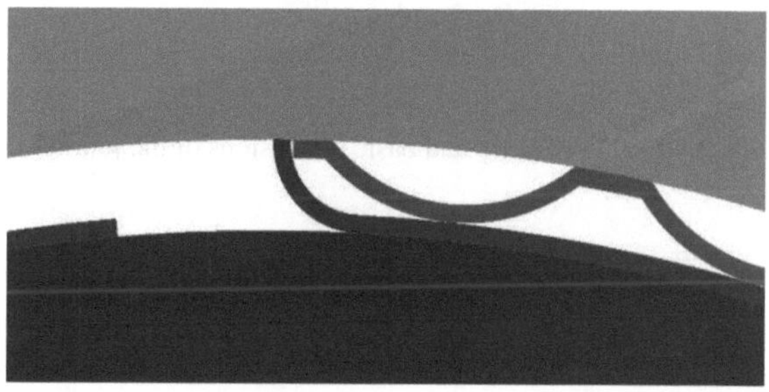

Fig. 3.11 Method of connecting and mutual situating elements of the foil bearing

For detailed studies to be possible of the phenomena occurring in foil bearings, a 3D model was developed, which does not have any simplifications with regard to the geometry of the bearing. For obvious reasons, only derogations in the geometry of a real bearing could not be taken into account from the assumed nominal shapes, which resulted from the performance inaccuracies and deformations of elastic thin, metal elements. This mainly concerned bump foil, which each bulging can slightly vary from others.

The 3D model consisted of four basic parts: the journal, sleeve, top foil and bump foil (Figs. 3.10 and 3.11). Just as in a real bearing, the journal and sleeve were separated by an assembly of thin foils, made up of top foil (slide) and bump foil (supporting). In the operating foil bearing, the journal rests on top foil, which is supported by bump foil. The load from the journal on the sleeve is transferred mainly by fragments of bump foil between the bumps. The attachment of both foils to the sleeve has been shown in Fig. 3.11. In an actual bearing, attachment is done by making a notch in the sleeve, into which the ends of both foils are placed and mounted using a threaded connection. The attachment of both foils using bonding technology is also encountered. The remainder part of the two foils remains free, and its displacements are limited by the surfaces of the journal and sleeve.

According to the drawings presented above, the developed model corresponded to the construction of the foil bearing of the 1st generation. That same model also allows to create the geometry of foil bearings of the 2nd and 3rd generation. This issue will be further explained later in this chapter, when discussing parameterization of the model of bump foil. In developing the model of this foil, it was assumed that its bumps are achieved through plastic processing using a punch and

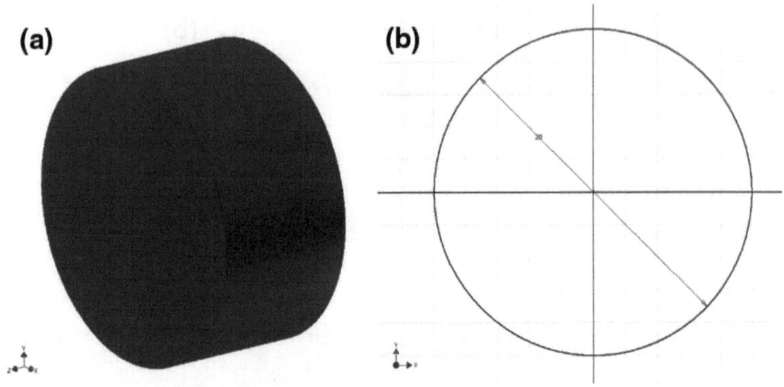

Fig. 3.12 3D geometry and basic dimensions of the journal

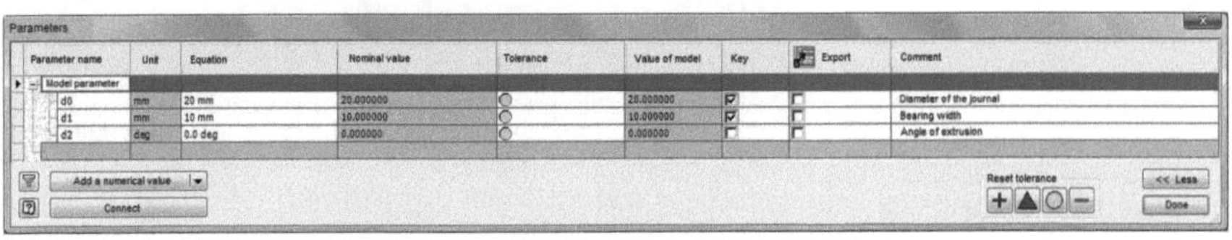

Fig. 3.13 Parameters specifying the dimensions of the journal

matrix. As a result of this process, a developed bump foil is achieved with bumps of the desired geometry, which target round shape is obtained after bending at a radius equal to the radius of the bearing.

Due to the possibility of a rapid reconstruction of the model and adjusting the geometry of particular elements of the bearing to the current study needs, a parameterized model was developed. In developing the sketches used to create the geometry of all parts of the foil bearing, elements of the sketches were fully dimensioned. Then a table with parameters and descriptions was created, so that if it is necessary to introduce changes to the dimensions, it took place by changing the parameter values, rather than directly at the level of the sketches. Such an approach, on the one hand, does not require from the user good knowledge of the program for designing, and on the other, better protects the sketches against improper operations, which may lead to their irreversible distortion and damage.

The 3D geometry of the bearing journal and basic sketch have been presented in Fig. 3.12. Figure 3.13 shows a table of the parameters used to fully define the geometry of this part.

For this part, the basic parameters were: journal diameter (d0) and bearing width (d1), which specifies the stretching length in the 3D model. In addition, Autodesk Inventor allows to determine the angle of convergence of stretching, which in the case of foil bearings will not be applied. All the relevant parameters, which can be changed by the user, have been marked as crucial. The remaining parameters when defining the geometry should remain unchanged, as they are treated as auxiliary.

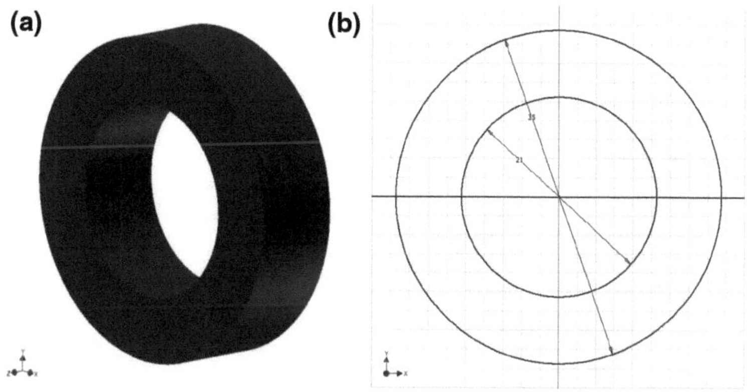

Fig. 3.14 3D geometry and basic dimensions of the sleeve

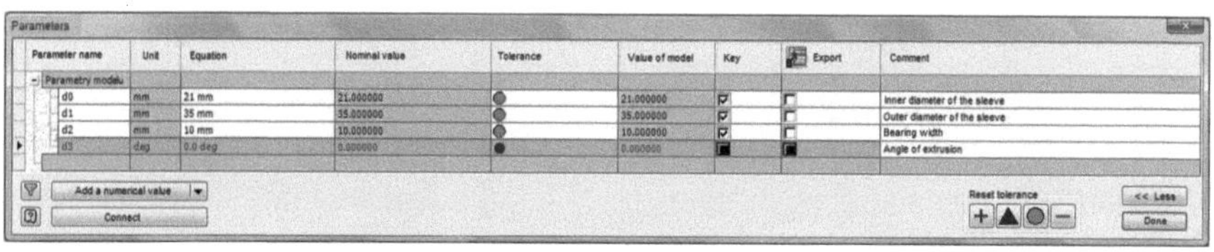

Fig. 3.15 Parameters determining the dimensions of the sleeves

Figure 3.14 shows the geometry and basic sketch of the bearing sleeve. It was assumed that the developed model shall consist of a full sleeve (one-piece) without elements attaching the assembly of foils and without a bearing rack. The structural solutions of these mountings are in fact an individual matter and it was difficult to propose a universal solution. The basic parameters determining the dimensions of the bearing sleeve were: internal diameter (d0), outer diameter (d1) and sleeve width (d2). Furthermore, as in the case of the journal, it was possible to change the convergence of stretching the sketch (Fig. 3.15).

Another element of the foil bearing was top foil (Fig. 3.16). In the case of this part, it was necessary to use as many as five control parameters (Fig. 3.17). These parameters were: top foil thickness (d1), angle determining the place of ending the top foil (d3), top foil width (d4), internal diameter of top foil—nominal diameter of the bearing (d5), internal diameter of bearing sleeve (d6). Additionally in the sketch, parameters determining the radius of bending the top foil was applied, as well as the angle of deflection of stretching. It was assumed in the proposed structural solution that the bent end of the top foil ends at the surface of the bearing sleeve, and as it has been previously mentioned, the system of attaching the foils was not included. The default location of attaching the top foil was the top part of the sleeve, but this can be changed freely by rotating the top foil in the assembly model. The developed model allows also to create the geometry of a bearing

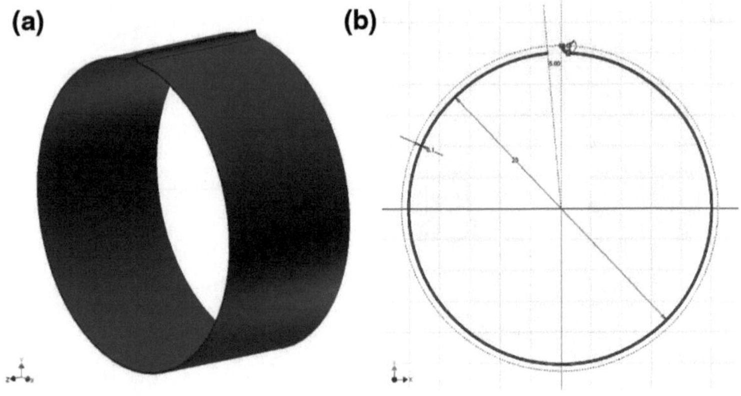

Fig. 3.16 3D geometry and basic dimensions of top foil

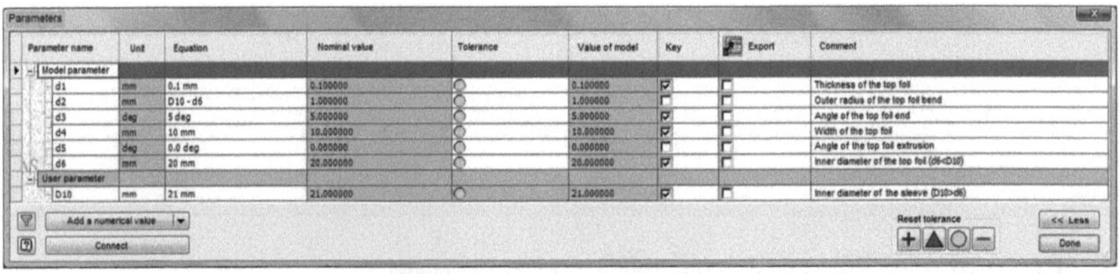

Parameters

Parameter name	Unit	Equation	Nominal value	Tolerance	Value of model	Key	Export	Comment
Model parameter								
d1	mm	0.1 mm	0.100000		0.100000	☑	☐	Thickness of the top foil
d2	mm	D10 - d6	1.000000		1.000000	☐	☐	Outer radius of the top foil bend
d3	deg	5 deg	5.000000		5.000000	☑	☐	Angle of the top foil end
d4	mm	10 mm	10.000000		10.000000	☑	☐	Width of the top foil
d5	deg	0.0 deg	0.000000		0.000000	☐	☐	Angle of the top foil extrusion
d6	mm	20 mm	20.000000		20.000000	☑	☐	Inner diameter of the top foil (d6<D10)
User parameter								
D10	mm	21 mm	21.000000		21.000000	☑	☐	Inner diameter of the sleeve (D10>d6)

Fig. 3.17 Parameters determining the dimensions of top foil

built from several top foils. By increasing the angle indicating the end of the top foil (d3) a model of the foil can be created which is only partly placed around the bearing journal.

The last and most complex part in geometry of the developed model was bump foil (Fig. 3.18). This foil consists of many bumps, which constitute the elasto-damping support of top foil. In order to clearly define the geometry of this part, it was necessary to apply up to 9 parameters. The dimensions of bump foil strictly depend on the dimensions of other elements of the foil bearing, therefore, it is recommended to begin modelling it after developing the remaining elements of the foil bearing.

The basic parameters determining the geometry of bump foil were: internal diameter of the bearing sleeve (d1), nominal diameter of bearing (d2), radius of bumps of bump foil (r1), distance of the beginning of bump foil from the attachment point of top foil (d6), angle determining the size of one segment of bump foil (k1), bump foil width (d8), number of bump foil bumps (d10), bump foil thickness (g1) and top foil thickness (g2). Additionally, the model had 6 auxiliary parameters visible in Fig. 3.19, which should not be changed by the user.

In order to develop a new geometry of bump foil, the following scheme should be followed. After introducing the diameter of the sleeve, foil and foil thickness, it is recommended to increase parameter k1, determining the size of one segment, which will enable changing the radius of bumps of the bump foil (r1) in a wider

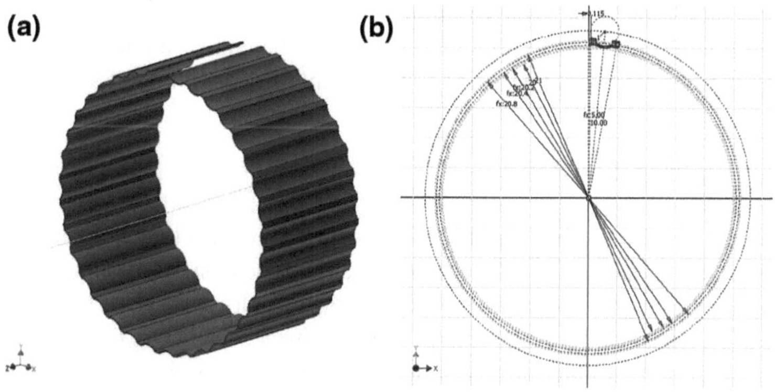

Fig. 3.18 3D geometry and basic dimensions of bump foil

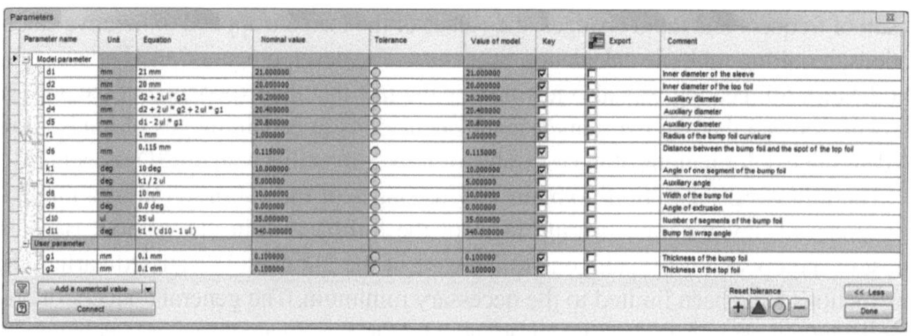

Parameter name	Unit	Equation	Nominal value	Tolerance	Value of model	Key	Export	Comment
Model parameter								
d1	mm	21 mm	21.000000	○	21.000000	☑	☐	Inner diameter of the sleeve
d2	mm	20 mm	20.000000	○	20.000000	☑	☐	Inner diameter of the top foil
d3	mm	d2 + 2 ul * g2	20.200000	○	20.200000	☐	☐	Auxiliary diameter
d4	mm	d2 + 2 ul * g2 + 2 ul * g1	20.400000	○	20.400000	☐	☐	Auxiliary diameter
d5	mm	d1 - 2 ul * g1	20.800000	○	20.800000	☐	☐	Auxiliary diameter
r1	mm	1 mm	1.000000	○	1.000000	☑	☐	Radius of the bump foil curvature
d6	mm	0.115 mm	0.115000	○	0.115000	☑	☐	Distance between the bump foil and the spot of the top foil
k1	deg	10 deg	10.000000	○	10.000000	☑	☐	Angle of one segment of the bump foil
k2	deg	k1 / 2 ul	5.000000	○	5.000000	☐	☐	Auxiliary angle
d8	mm	10 mm	10.000000	○	10.000000	☑	☐	Width of the bump foil
d9	deg	0.0 deg	0.000000	○	0.000000	☐	☐	Angle of extrusion
d10	ul	35 ul	35.000000	○	35.000000	☑	☐	Number of segments of the bump foil
d11	deg	k1 * (d10 - 1 ul)	340.000000	○	340.000000	☐	☐	Bump foil wrap angle
User parameter								
g1	mm	0.1 mm	0.100000	○	0.100000	☑	☐	Thickness of the bump foil
g2	mm	0.1 mm	0.100000	○	0.100000	☑	☐	Thickness of the top foil

Fig. 3.19 Parameters determining the dimensions of bump foil

range. Only after obtaining the desired geometry of bumps should the target value be provided of the angle determining the size of one segment (k1). Then, after selecting the distance defining the point of attaching the bump foil (d6), the value of the parameter determining the number of bumps should be entered (d10), so as to obtain bump foil of the desired length. The distance of attaching bump foil (d6) from the point of attachment of top foil should, in addition to the thickness of top foil, take into account its curvature, so that the foils do not overlap one another. If, when entering the new parameters, an error occurs resulting from trying to create flawed geometry, the program will automatically display a message about problems with updating the sketch. To return to the previous solution, click on the button "cancel" and then "undo". When entering changes to the parameters, it is recommended to perform updates by clicking the "finish" button after changing each single parameter.

The developed model of bump foil and the entire bearing also allows to develop the geometry of a foil bearing of the 2nd and 3rd generation. The dimensions of a single bump foil can in fact be limited by reducing the number of bumps (d10) and foil width (d8). In the assembly model, however, the inserted bump foil should be copied by using the command "circular pattern". If it is necessary to apply several different bump foils in the same bearing, it would be necessary, however, to copy the developed bump foil model and enter changes in the copied model. Then different variants of bump foil would need to be inserted into one assembly of a foil bearing.

3.3.1.1 Simplified 2-D Model

A 2D model of a foil bearing was developed mainly for conducting analyses while taking into account dynamic loads. This model is a simplification of a 3D model, and during its development it was assumed that due to the cylindrical shape of top foil and the method of its load, mainly displacements of this element occur along the circumference. The top foil moving within the sleeve does not deform along the bearing width. As a result of bends in the top foil it is very stiff in this direction, which has been confirmed by previously conducted studies [7, 12, 13]. 2D models in relation to 3D models consist of a much smaller number of degrees of freedom, which results in reducing the time of calculations. This is very important in the analysis of dynamic phenomena, where, as a result of frequent and rapid changes, the values of the load are applied to iterative calculations, requiring multiple calculation processes. The development of an alternative 2D model of a foil bearing was also justified by the fact that the experience of the authors thus far showed that the two-dimensional calculations in certain ranges of loads allow to determine deformations of the assembly of foils compatible with the 3D model [14].

The following subsections explain various stages of the construction of a numerical model of the structural bearing layer of the foil bearing in 2D version. Because most of the operation is analogous to the case of the 3D model discussed earlier, the instructions have been limited to the necessary minimum. The general concept of creating a model using Autodesk Inventor and ABAQUS software remained unchanged.

The 2D model reflected the nominal geometry of foil bearings in its transverse cross-section. As in the case of the 3D model, it consisted of four parts: the journal, bump foil, top foil, and sleeve. This time, instead of building the 3D geometry using stretching based on sketches, the option "close envelope" was applied. In this way, 2D models of elements of the foil bearing were created. The realized bearing model constituted a transverse cross-section of bearings of an infinitely small thickness. It has been presented in Fig. 3.20.

Fig. 3.20 The geometry of a 2D model of a foil bearing—assembly diagram

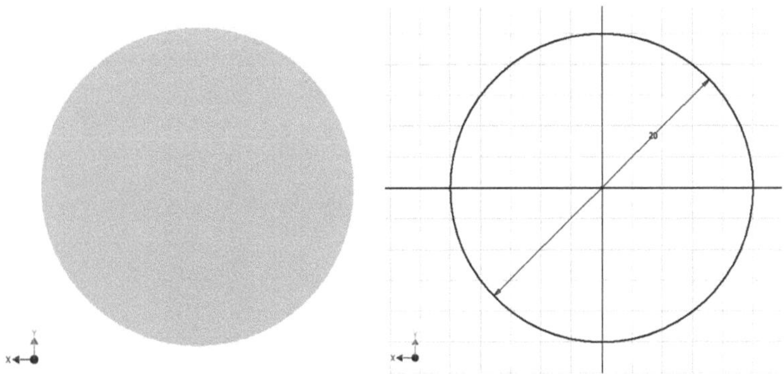

Fig. 3.21 2D geometry and basic dimensions of the journal

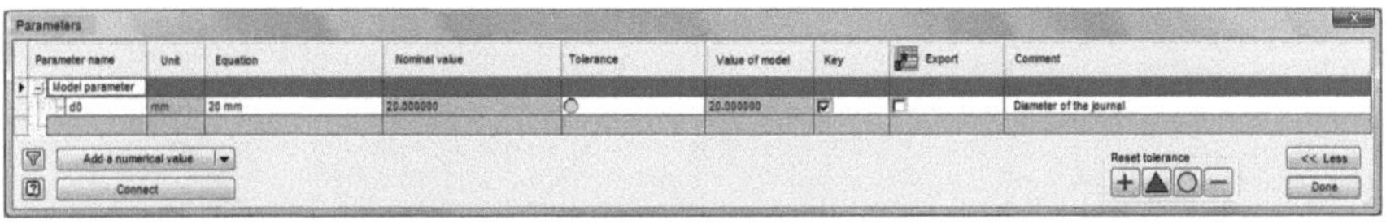

Fig. 3.22 Parameters specifying the dimensions of the journal

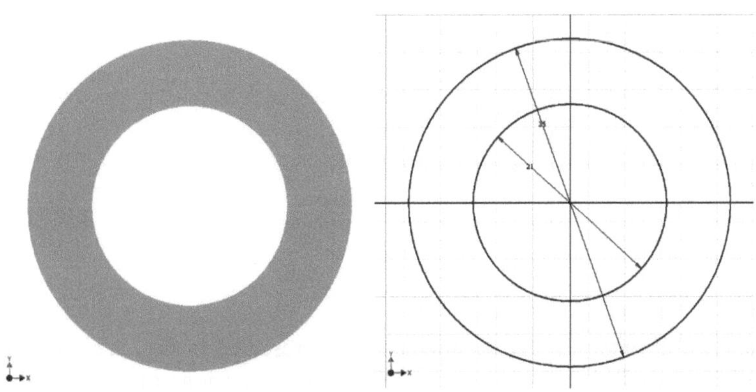

Fig. 3.23 2D geometry and basic dimensions of the sleeve

The parameterization of the 2D geometric model proceeded in the same way as for a 3D model. In the case of every part of the bearing, 2 parameters were useless, specifying the bearing width (length of stretching) and the convergence of stretching. The 2D model of a journal and its parameters have been presented in Figs. 3.21 and 3.22.

Figure 3.23 shows the geometry of the foil bearing sleeve. The parameters controlling the dimensions of the sleeves have been shown in Fig. 3.24.

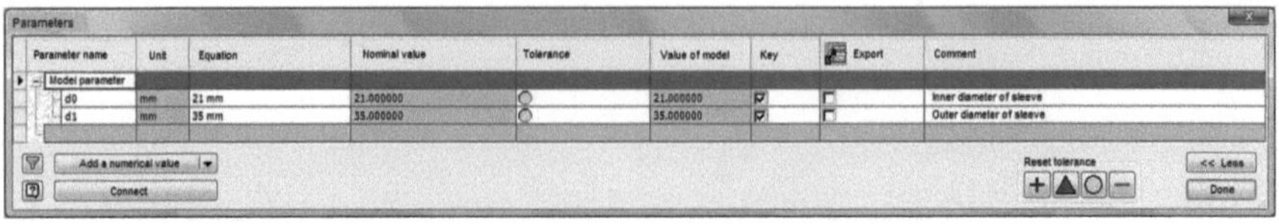

Parameters

Parameter name	Unit	Equation	Nominal value	Tolerance	Value of model	Key	Export	Comment
Model parameter								
d0	mm	21 mm	21.000000		21.000000			Inner diameter of sleeve
d1	mm	35 mm	35.000000		35.000000			Outer diameter of sleeve

Fig. 3.24 Parameters determining the dimensions of the sleeves

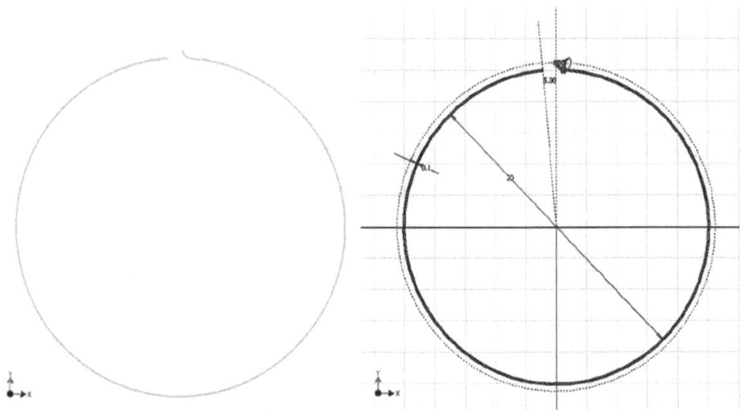

Fig. 3.25 2D geometry and basic dimensions of top foil

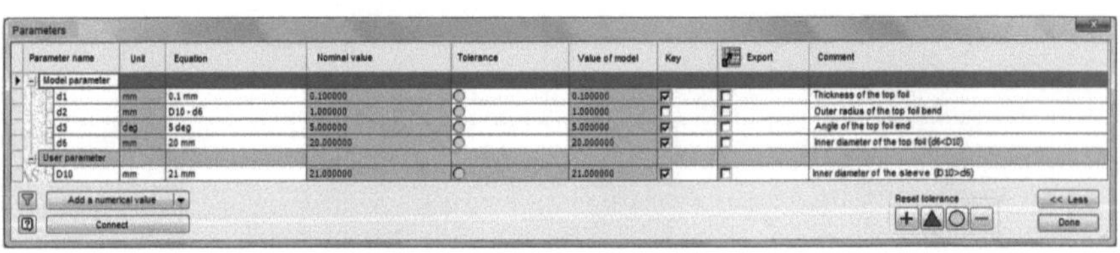

Parameters

Parameter name	Unit	Equation	Nominal value	Tolerance	Value of model	Key	Export	Comment
Model parameter								
d1	mm	0.1 mm	0.100000		0.100000			Thickness of the top foil
d2	mm	D10 - d6	1.000000		1.000000			Outer radius of the top foil bend
d3	deg	5 deg	5.000000		5.000000			Angle of the top foil end
d6	mm	20 mm	20.000000		20.000000			Inner diameter of the top foil (d6<D10)
User parameter								
D10	mm	21 mm	21.000000		21.000000			Inner diameter of the sleeve (D10>d6)

Fig. 3.26 Parameters determining the dimensions of top foil

The top foil model in 2D version has been presented in Fig. 3.25. The table in Fig. 3.26 shows the parameters controlling the dimensions of foils. Both in the case of the 2D model and the previously discussed 3D model, the internal diameter of top foil (nominal diameter of bearing) is by default equal to the diameter of the bearing journal. In the actual foil bearings the clearance between top foil and the journal is selected by taking into account the operational conditions of the bearing. It depends on many factors, such as: top layer material of the journal and top foil, nominal rotational speed of the machine, type of machine and geometry of the fluid-flow system, or the weight of the shaft. The most common structural solution of foil bearings, in which the top foil is initially clamped on the journal, and the fluid-flow layer of the lubricant medium is formed only when a certain speed is achieved.

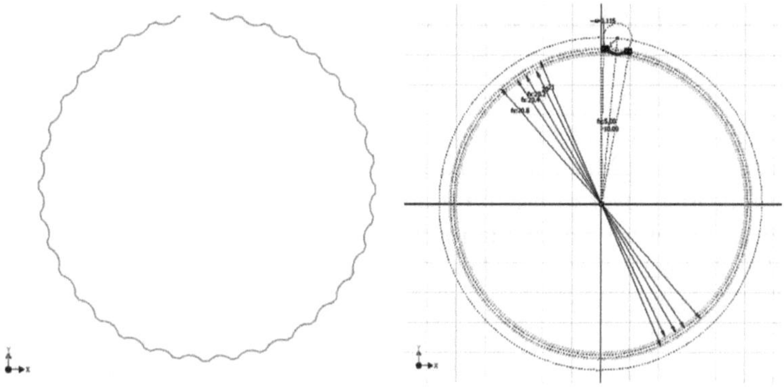

Fig. 3.27 2D geometry and basic dimensions of bump foil

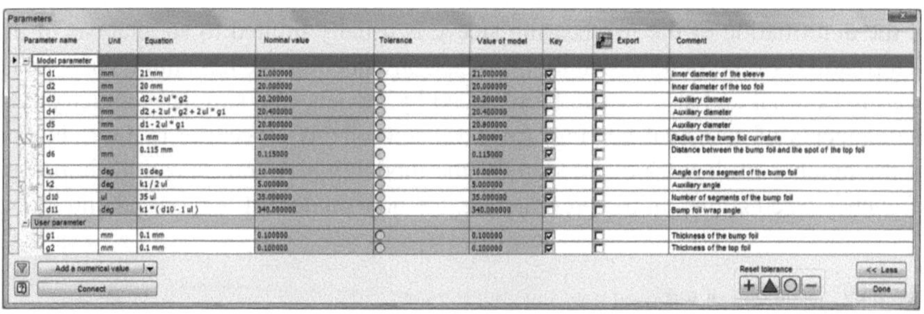

Parameters

Parameter name	Unit	Equation	Nominal value	Tolerance	Value of model	Key	Export	Comment
Model parameter								
d1	mm	21 mm	21.000000		21.000000	✓		Inner diameter of the sleeve
d2	mm	20 mm	20.000000		20.000000	✓		Inner diameter of the top foil
d3	mm	d2 + 2 ul * g2	20.200000		20.200000			Auxiliary diameter
d4	mm	d2 + 2 ul * g2 + 2 ul * g1	20.400000		20.400000			Auxiliary diameter
d5	mm	d1 - 2 ul * g1	20.800000		20.800000			Auxiliary diameter
r1	mm	1 mm	1.000000		1.000000	✓		Radius of the bump foil curvature
d6	mm	0.115 mm	0.115000		0.115000	✓		Distance between the bump foil and the spot of the top foil
k1	deg	10 deg	10.000000		10.000000	✓		Angle of one segment of the bump foil
k2	deg	k1 / 2 ul	5.000000		5.000000			Auxiliary angle
d10	ul	35 ul	35.000000		35.000000	✓		Number of segments of the bump foil
d11	deg	k1 * (d10 - 1 ul)	340.000000		340.000000			Bump foil wrap angle
User parameter								
g1	mm	0.1 mm	0.100000		0.100000	✓		Thickness of the bump foil
g2	mm	0.1 mm	0.100000		0.100000	✓		Thickness of the top foil

Add a numerical value

Connect

Reset tolerance

<< Less

Done

Fig. 3.28 Parameters determining the dimensions of bump foil

Figure 3.27 shows the model of bump foil. The final geometry consisting of dozens of bumps was obtained as a result of duplicating the base bumps in a circular pattern. Then, it was necessary to combine all sectors of the bump foil so that they form one part. In the process of making changes to the geometry of the bump foil it is recommended to use the same rules as in the case of the 3D model. It is very important that before modifying the radius of the bump foil's bumps, the value of the angle determining the size of one segment is increased. This will extend the possibility of changing the value of the radius of bumps. In the event of a conflict occurring in the dimensions, the performed operations should be cancelled, which will allow to restore the state of the model before the modifications (Fig. 3.28).

The 2D model of the geometry of the foil bearing was developed to minimize the time necessary to conduct analyses, in particular dynamic phenomena. A total of two alternative models of the structural bearing layer were created, enabling an accurate and detailed analysis in the range of static loads (3D model) and a less accurate, but very efficient analysis in the range of dynamic loads (2D model).

3.3.2 Numerical Model of the Foil Bearing Structure

In order to take advantage of the developed model of the geometry of a foil bearing, for numerical calculations, the discretization of geometry was necessary and the creation of a FEM model on its basis. After developing the geometry of the structural bearing layer of the foil bearing in the program Autodesk Inventor, the file of assembly was saved in the format (*.sat), which allowed it to be opened in ABAQUS. The next step was to load the file of the model in ABAQUS. After these steps, the program converts the geometry of all parts of the model. The following diagram (Fig. 3.29) shows the geometry of a top foil displayed in ABAQUS. The additional breakdown of the foil model into smaller fragments is visible in the diagram, which was introduced in order to prepare for discretization.

Figure 3.30 shows the geometry of the bump foil model, which was imported into ABAQUS. Because the assembly of sensitive foils constitutes a system characterized by a much lower stiffness than the bearing sleeve and journal [12], the analysis of the deformation of the foil bearing sleeve is usually limited to an analysis of the deformations of the same assembly of foils. Consequently, the remaining parts of the bearing (journal and sleeve) can be replaced by models of a simpler geometry than the elements of an actual bearing. An advantage of this procedure is the reduction of the number of degrees of freedom of the model, which translates into improved efficiency calculations. In the present case, the new models of the journal and sleeve

Fig. 3.29 Geometry of top foil imported into ABAQUS

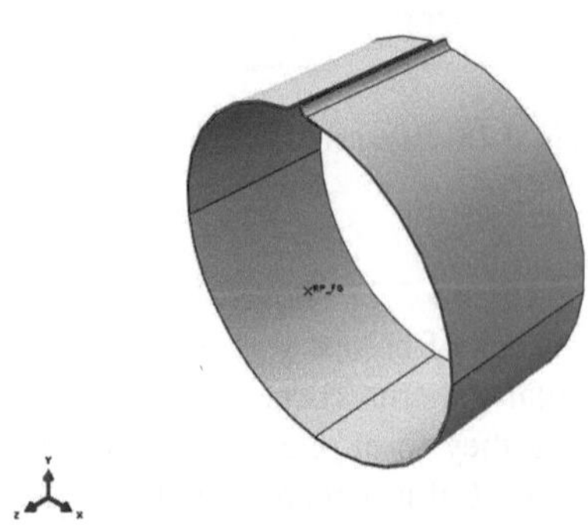

Fig. 3.30 Geometry of bump foil imported into ABAQUS

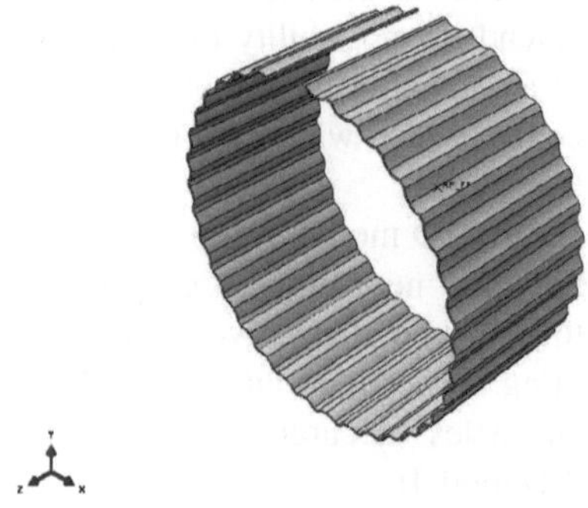

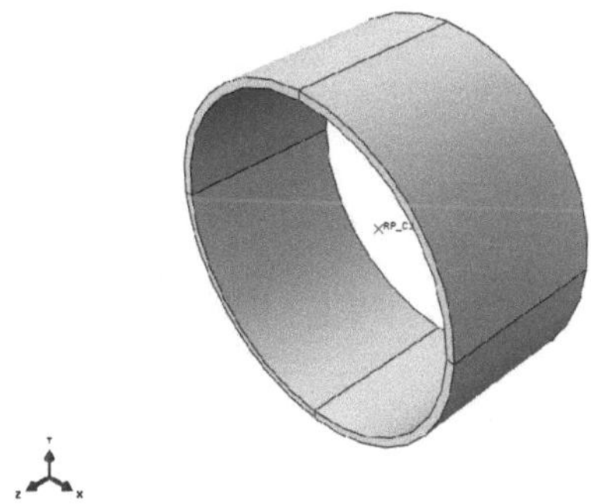

Fig. 3.31 Geometry of a rigid journal developed in ABAQUS

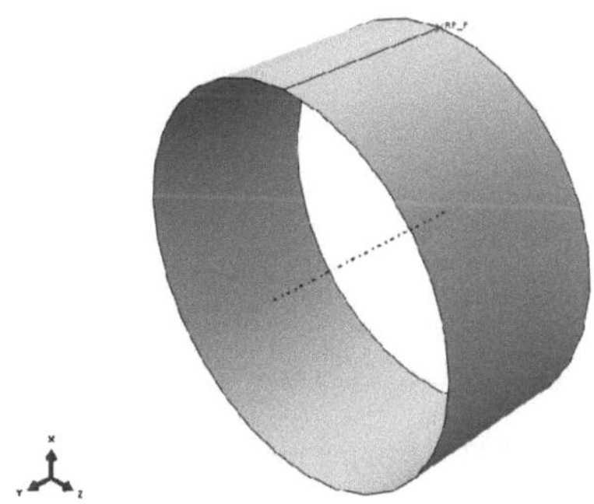

Fig. 3.32 Geometry of a rigid sleeve developed in ABAQUS

were developed directly in ABAQUS and have been presented in Figs. 3.31 and 3.32. A characteristic feature of these models is the very simple geometry, which can be easily developed using simple tools provided for modelling in ABAQUS.

Elements of the foil bearing not used for the analysis of deformations of the assembly of foils developed using Autodesk Inventor (journal and sleeve) could in fact be used for other types of analyses. An example might be the analysis of thermal phenomena, where to include heat flow in all parts of the elements of the bearing node, it would be necessary to take into account the full geometry of the sleeve and journal. A similar situation would take place in the analysis of the sensitivity of the dynamic bearing support [15].

After loading the models of individual parts and developing the geometry of the missing items, the properties of construction materials should be defined (Fig. 3.33). In this case, it was assumed that the elements of the structural bearing layer are made of steel with the following properties: $\rho = 7{,}860\ \mathrm{kg/m^3}$, $E = 2.1 \times 10^{11}$ Pa, $\nu = 0.3$. The next step is to create a new solid section and assign the relevant properties to all the foil bearing elements. This is done by assigning previously created sections to the bearing elements (Fig. 3.33). Since in the present case the new model of the bearing sleeve has been developed as a rigid part, it does not require defining of the material.

The fully defined parts of the foil bearing should be assembled together using the tools provided for this. The proper use of these tools will allow to assemble the elements as shown in Fig. 3.34, which presents the model of the structural bearing layer of the foil bearing along with the rigid journal.

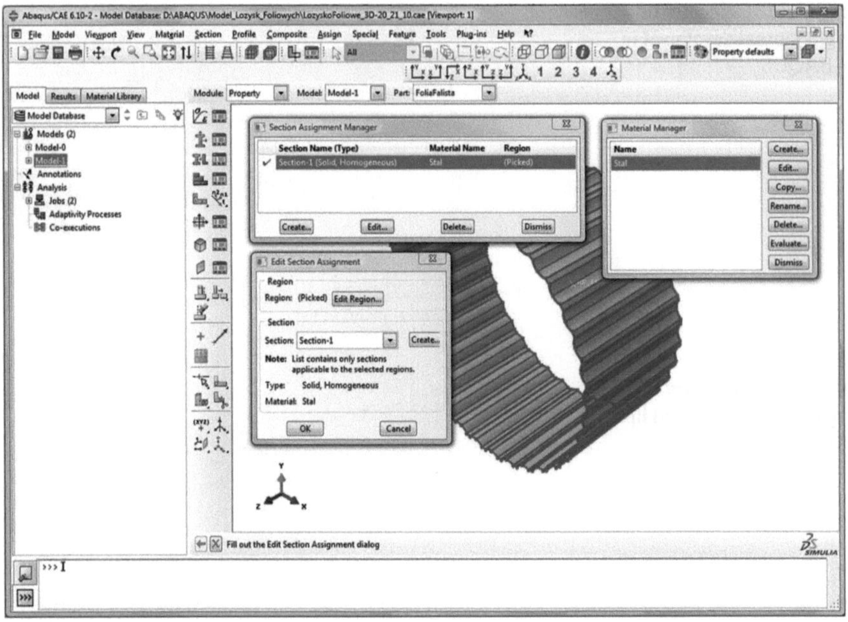

Fig. 3.33 Defining material properties in ABAQUS

Fig. 3.34 Model of foil bearings after assembling all parts together

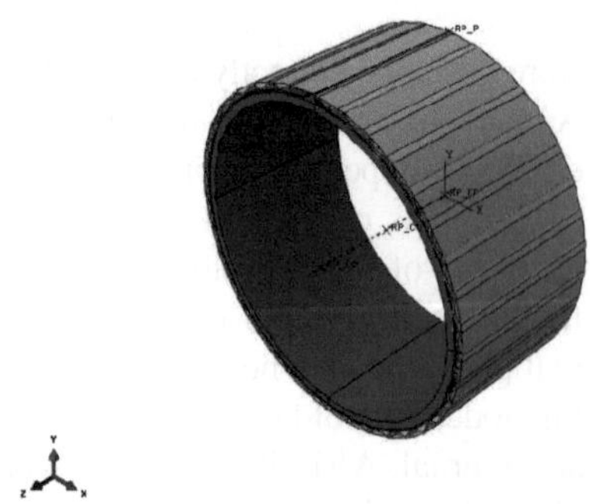

The subsequent stages of developing the numerical model depend greatly on the type of the selected analysis. Figure 3.35 shows the parameters used in the static analysis. Also the parameters of contact need to be defined between all the parts of the model cooperating with one another: the journal and top foil, top foil and bump foil and bump foil and sleeve (Fig. 3.36). As the output value, the "surface-to-surface" type contact should be selected, and in the contact parameters, the friction factor should be entered, e.g. with a value of 0.1.

The next stage of developing the FEM model is to define the loads in the module. Discussed below are two cases of loading the assembly of foils. In the first case, the foils were subjected to a concentrated force of 200 N (pointing downwards vertically) through the journal (Fig. 3.37). The case corresponds to the work of the bearing at lower rotational speeds, before a fluid-flow carrying layer

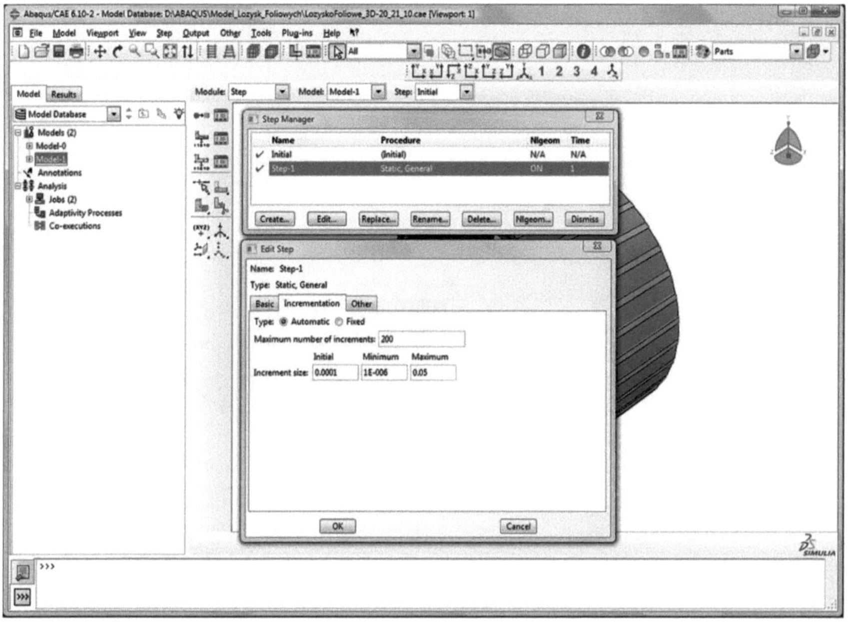

Fig. 3.35 Defining the type and parameters of the analysis in ABAQUS

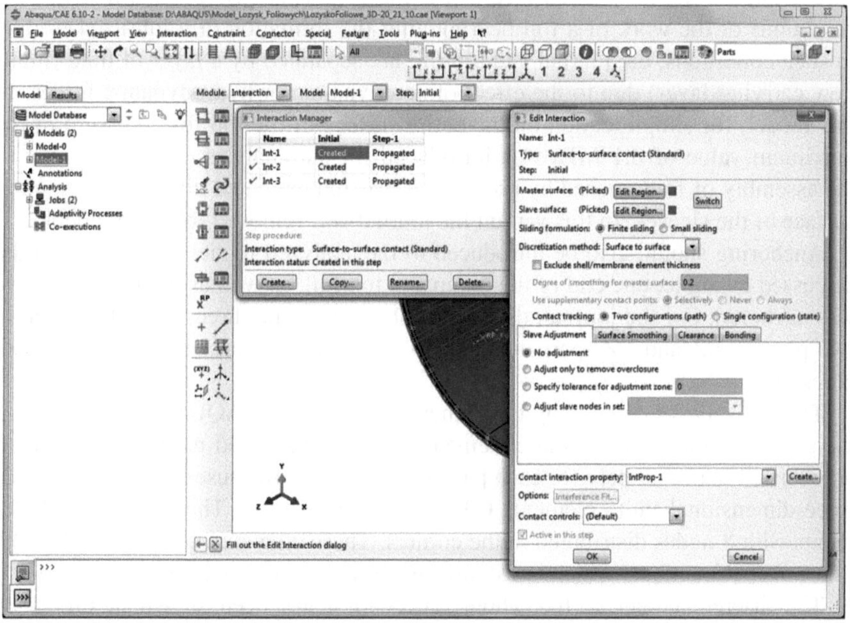

Fig. 3.36 Defining the contact between elements of the foil bearing in ABAQUS

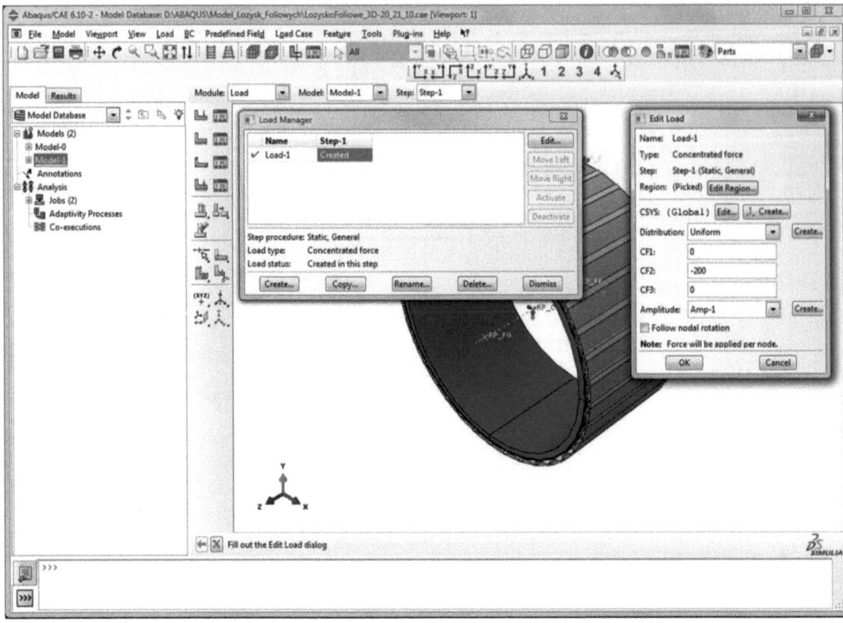

Fig. 3.37 Defining loads in ABAQUS

is formed. In the second analyzed case, the assembly of foils was loaded by pressure acting on the internal surface of the top foil. This case of load matches the conditions of the work of a foil bearing after the formation of a lubricant wedge. In such conditions, the journal and top foil are separated by a layer of fluid (fluid-flow carrying layer) due to the effect of aerodynamic or hydrodynamic forces. In this model, for simplification on the entire surface of the top foil, pressure of the maximum value of 0.1 MPa was introduced. In the analysis of deformations of the assembly of foils caused by pressure, it was implemented directly on the inner surface of the sleeves. In this variant the journal was removed from the model.

Anchoring should also be introduced to the analyzed model (Fig. 3.38). In the discussed example, one end of the bump and top foil, as well as the bearing sleeve was anchored. For the journal, displacements in the axial direction and rotation in two planes (XZ and YZ) perpendicular to the plane of rotation of the rotor were blocked.

The last stage of developing the numerical model in ABAQUS is the division of the geometry into discrete finite elements. For the discussed mode, discretization was carried out separately for each part of the bearing. Because the model was 3D, three-dimensional finite elements C3D8R [16] were used. These are cuboid elements with 8 nodes distributed in the corners. These elements are characterized by linear shape functions, which is recommended for contact analyses.

The discrete model of the whole foil bearing has been shown in Fig. 3.39. Figure 3.40 shows the discretization of top and bump foil. To divide the top foil,

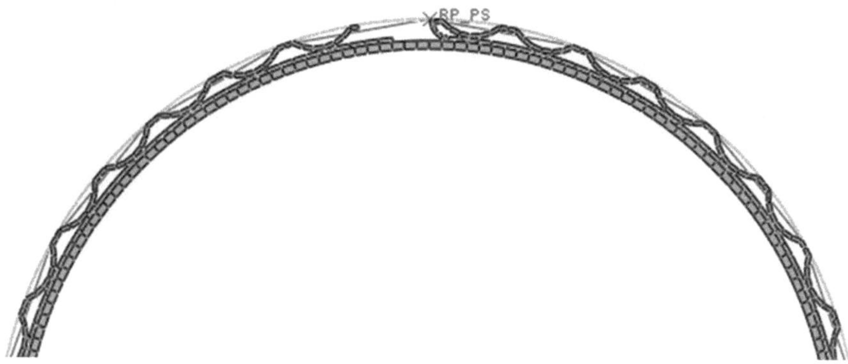

Fig. 3.38 Defining boundary conditions in ABAQUS

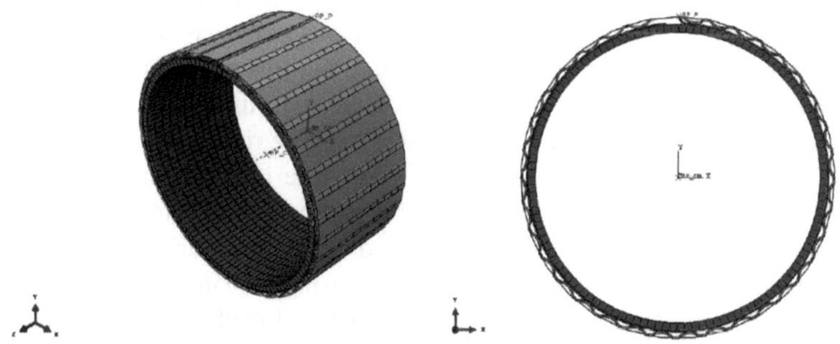

Fig. 3.39 Discretization of the foil bearing model—assembly

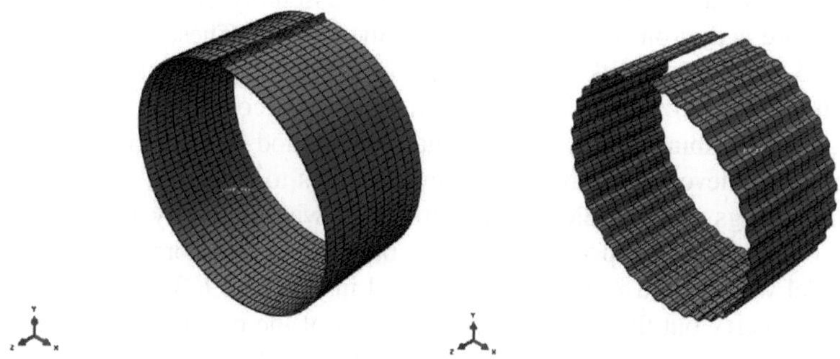

Fig. 3.40 Discretization of the foil bearing model—top foil and bump foil

1524 C3D8R elements were used. In order to project the geometry of bump foil, it was necessary to use up to 2112 elements. To divide the bearing journal (Fig. 3.41) the same finite elements C3D8R in the number of 1,488, however, in the case of this part, the additional option "Rigid body" was applied, which causes the discretized

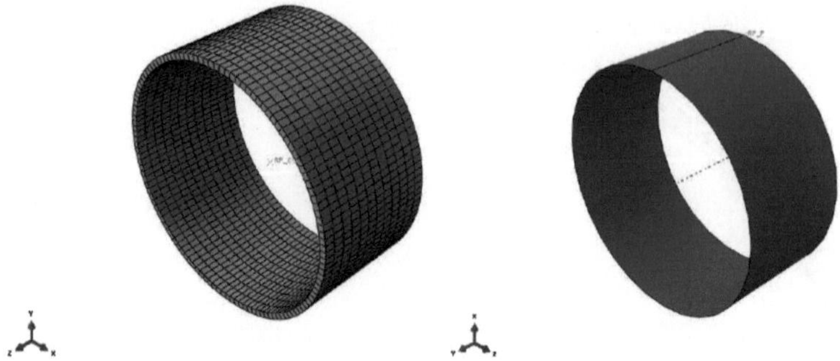

Fig. 3.41 Discretization of the foil bearing model—rigid journal and sleeve

element to be treated as a rigid body. The force applied to one point of this element is then transferred by the whole part. Finite elements in the case of "Rigid body" are non-deformable, however, their geometry is used during contact analysis.

Because for the needs of the analysis of the deformation of the assembly of foils, the bearing sleeve was developed as an "Analytic rigid shell", this part did not require division into finite elements (Fig. 3.41). The option "Analytic rigid shell" was not used in the case of the journal, as this part cannot transfer loads in the form of concentrated force, but can only be subjected to kinematic drive. The entire model of the structure of the foil bearing was described using 86,058 degrees of freedom.

In order to validate the correctness of the developed numerical model after developing the FEM mesh, a trial analysis of the deformation of the assembly of foils was launched for two variants of loads. The first variant of the load consisted in concentrated force acting on the assembly of foils through the journal. In the second variant, the deformations of the assembly of foils were caused by the pressure acting on the inner surface of the top foil. These calculations above all aimed to initially check the correctness of the developed geometry, the applied method of discretization of the model, boundary conditions, as well as the method of describing contact phenomena. The method of discretization discussed above was achieved after attempting several times to carry out calculations with various settings of the FEM mesh. Initially, it was necessary to make changes in the originally developed FEM model, but after a few corrections, a numerical model was created in the form presented in Figs. 3.39, 3.40 and 3.41, which allowed to carry out the simulation. The results of the calculations for the analysis, at loading the assembly of foils with concentrated force through a rigid bearing journal have been presented in Fig. 3.42. This diagram clearly demonstrates that, in accordance with the expectations, the largest stresses occurred in the areas of a few lower bumps of bump foil, which transferred load to the greatest extent. At the introduced maximum load, the reduced stresses slightly exceeded 70 MPa.

At loading the tested system with pressure, the largest stresses occurred in the fold area of top foil, at its anchored end. This is shown in Fig. 3.43. The greatest displacements of the assembly of foils occurred at the free, unsupported end of the top foil.

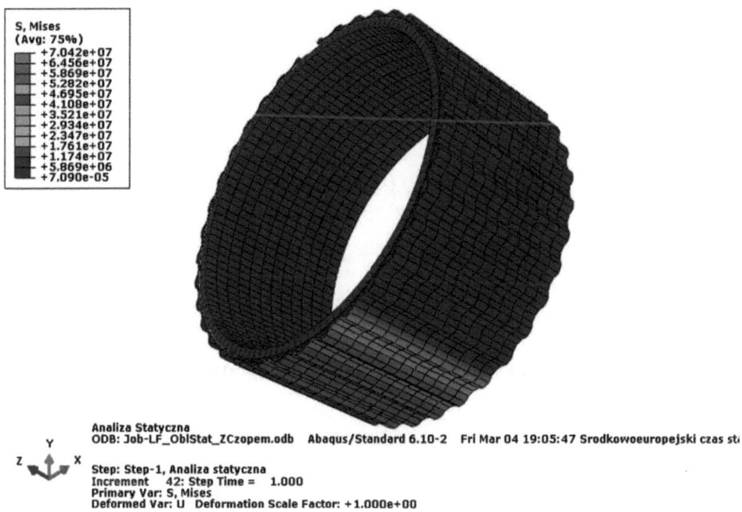

Fig. 3.42 Results of the calculations at loading the assembly of foils with concentrated force

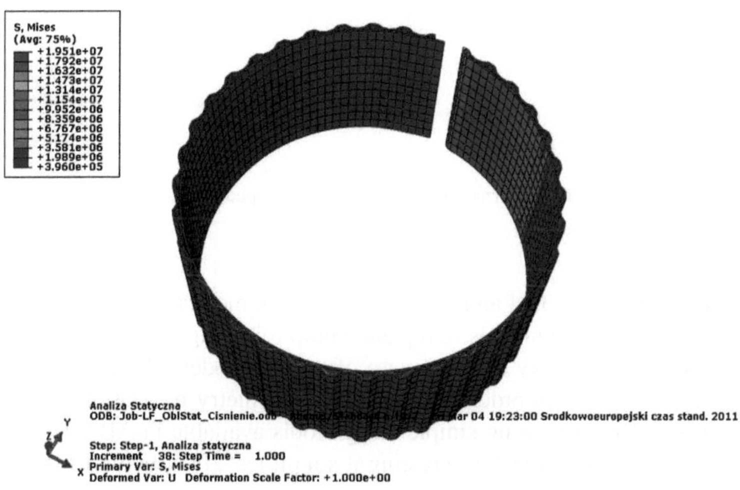

Fig. 3.43 Results of the calculations at loading the assembly of foils with pressure

3.3.2.1 Simplified 2-D Model

The different stages of developing the FEM model in 2D, were carried out in the same order as for the previously discussed 3D variant. Below, the discussion focuses on the most important operations resulting from applying a 2D description.

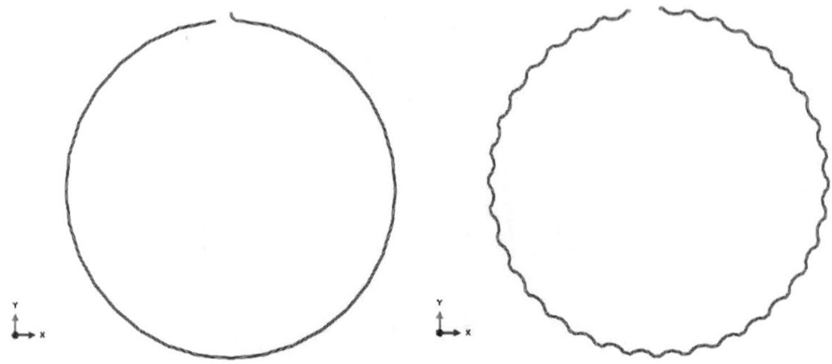

Fig. 3.44 2D geometry of the model of top and bump foil imported into ABAQUS

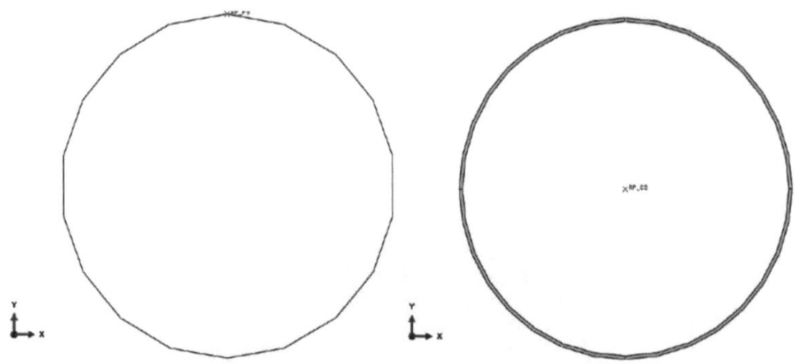

Fig. 3.45 2D geometry of the journal and rigid sleeve developed in ABAQUS

The drawings (Figs. 3.44 and 3.45) present the numerical models of all parts of the 2D foil bearing. The models of top and bump foils (Fig. 3.44) were loaded into ABAQUS in the same way as in the case of the 3D model. Models of the journal and sleeve (Fig. 3.45), in order to simplify the geometry in relation to the actual system, were developed using simple design tools available in ABAQUS. Various models were compiled together, creating a simplified 2D structural model of the entire foil bearing (Figs. 3.46 and 3.47).

In order to assign the 2D model a specific width, necessary to properly calculated the input and result data, in the module “Property”, the width of the analyzed bearing should be defined. This is done by entering the appropriate values (in the present case, in meters) in the “Edit section” window, which appears after creating a new section. The width of the bearing is entered in the field “Plane stress/strain thickness”. All the results of the strength analysis will include the provided width value of such a defined system.

Due to the fact that for a 2D geometry, it is possible to apply only two-dimensional finite elements, for the discretization of top foil, bump foil and the journal, CPE4R type elements were used [16]. These are rectangular elements with four nodes and functions

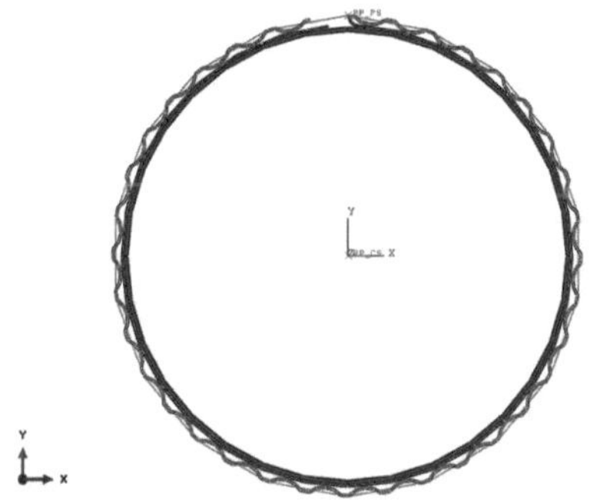

Fig. 3.46 2D model of the foil bearing after assembling all parts together

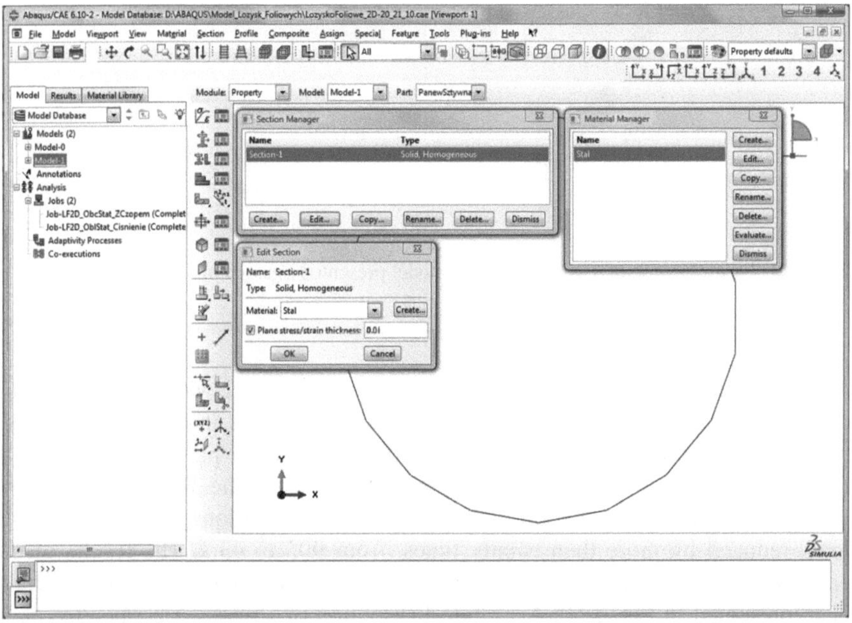

Fig. 3.47 Defining material properties and widths of the bearing in ABAQUS

of the shape of the first stage. The bearing journal, however, was treated as a rigid body. The option "Rigid body" served this. In this model, the sleeve was developed as a 2D rigid element "2D Discrete Rigid." Therefore, for the division of this part, the only possible elements in this case, type R2D2 were used [16]. The model consisted of a total of 6,354 degrees of freedom. All the parts of the foil bearing, after dividing into finite elements, have been presented after assembly in Fig. 3.48.

For a 2D model, in accordance with the procedures previously adopted, calculations were done for two variants of static load in the form of concentrated force acting on the journal and pressure loading the sliding foil. For such conditions, calculation files were created and an analysis was started.

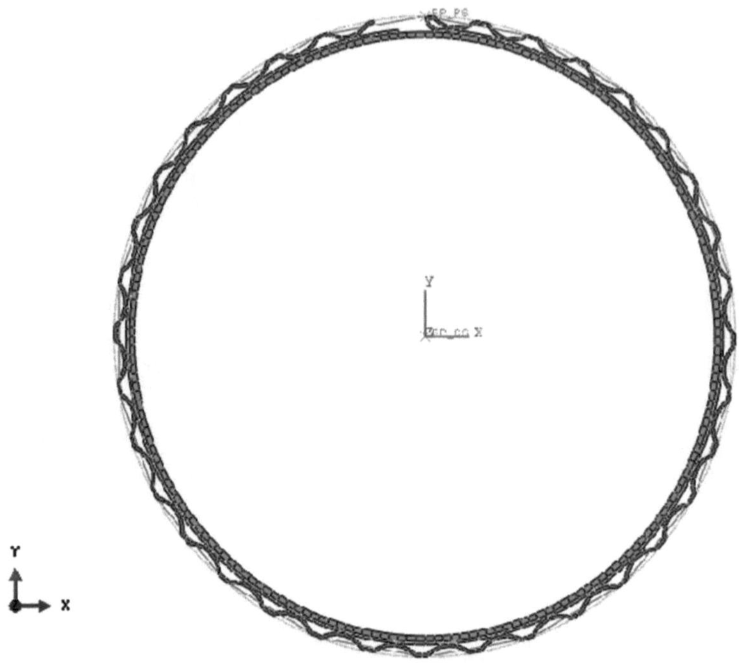

Fig. 3.48 Discretization of a 2D foil bearing model presented in assembly

The results of the calculations obtained for the model with the journal and concentrated force have been presented in Fig. 3.49. Like in the case of the 3D model, the largest values of stresses occurred in the lower part of the bump foil. Furthermore, they achieved similar values as when using the 3D model. The time needed to conduct calculations in corresponding conditions of anchoring and load, reduced by more than twenty times, from 905 to 43 s. The results of the calculations in the form of reduced stresses for a system loaded by pressure have been presented in Fig. 3.50. A similar distribution of stresses was obtained as in the case of the 3D model, as the largest stresses occurred in the attachment area of the top foil. However, there were differences in the values of stresses, as in the case of the 2D model, they were higher by approx. 50 %. By using the 2D model, an almost 40-fold reduction in the calculation time was achieved from 587 to 15 s.

According to preliminary results of the analysis, the use of a simplified 2D model of foil bearings has a very positive effect on reducing the time required to perform the calculations. In these cases, this time was reduced by a few dozen times. However, the simplification of the model took place at the cost of accuracy of the obtained results. The solution to this problem can be a detailed verification and tuning of the model by using the results of experimental tests. These issues are the subject of the following subsections of this chapter.

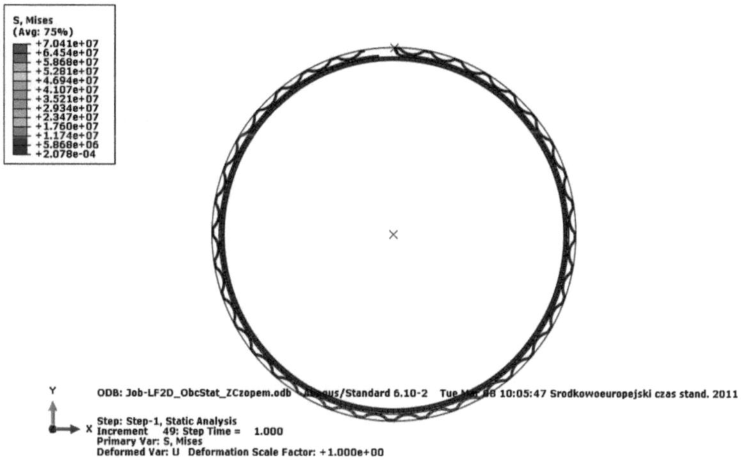

Fig. 3.49 Results of the calculations at loading the assembly of foils with a force through the journal (2D model)

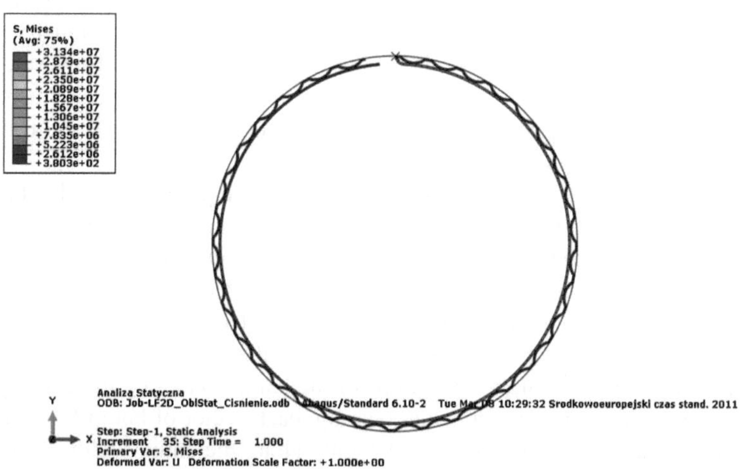

Fig. 3.50 Results of the calculations at loading the assembly of foils with pressure

3.3.3 Verification of the Structural Model

This subsection discusses the next stage of work on the numerical model of the foil bearing. As part of this stage, experimental verification and tuning of the model was carried out. These tests were divided into two stages, differing in the way the system was loaded.

Table 3.3 Basic dimensions and parameters of the tested foil bearing

No.	Dimension/parameter	Value
1	Internal diameter	38.17 mm
2	Bearing length	38.10 mm
3	Nominal diameter of the journal	38.10 mm
4	Nominal (project) clearance	0.035 mm
5	Number of bumps of bump foil	25
6	Pitch of bump foil	4.57 mm
7	Length of bumps	4.06 mm
8	Height of bumps	0.38 mm
9	Foil thickness	0.1 mm
10	Poisson number	0.29
11	Young's modulus	2.1×10^{11} Pa

3.3.3.1 Verification in the Range of Static Loads

The different stages of developing the geometry of the foil bearing were discussed in detail in the previous subsection of this chapter. Due to the characteristic for 3D models low calculation performance, these studies were based on a simplified, 2D geometry of the foil bearing. This model fully reflected the geometry of the tested bearing in the plane perpendicular to the axis of rotation. However, it did not include the change in the shape along the width of the bearing. By using this model, it was also possible to project the structural bearing layer of the foil bearing of the 1st and 2nd generation.

The basic dimensions of the tested bearing were adopted on the basis of literature [17], so that it was possible to compare the results of the simulation with the test results obtained by other researchers. The adopted dimensions and other parameters of the bearing have been summarized in Table 3.3.

The bump foil of an actual bearing, on the basis of which the model was developed, consisted of five same sectors of foils distributed equally along the perimeter of the sleeve. The total number of bumps of the bump foil along the entire perimeter was 25. Additionally, in an actual bearing, each sector of bump foil consisted of the same four narrow sheets, spaced evenly along the entire width of the bearing. The division of bump foil sectors into four smaller foils was not projected in the model due to its two-dimensionality. Because the journal and sleeve of the foil bearing are elements of a much higher stiffness than thin foils, during analysis these two parts were treated as rigid solids.

The numerical model was developed in ABAQUS as discussed in the previous subsection. Simulation tests were planned in such a way as to project during the analyses the conditions of the experiment performed by American researchers [17]. During the simulation, the journal of the tested bearing was loaded by static force with a maximum value of 224 N. The value of the force increased linearly along with the time of the analysis, reaching maximum after 1 s. One end of the top foil and one end of each fragment of the bump foil has been attached to the

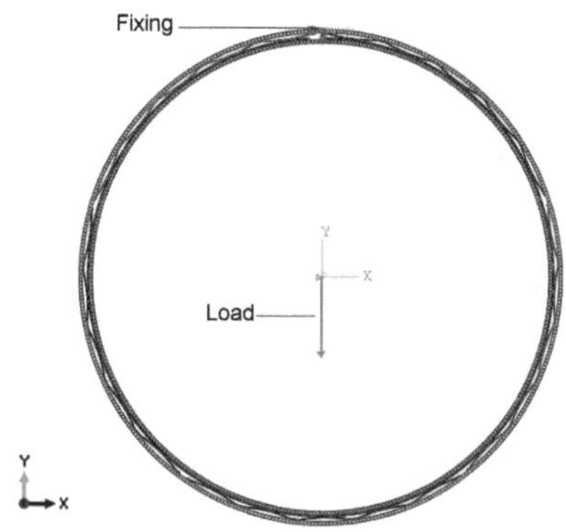

Fig. 3.51 FEM model of the foil bearing with anchoring and load

surface of the sleeve. The displacements of free fragments of the foil were limited by the surfaces of the journal and sleeves, between which the contact was modelled, with a friction coefficient of 0.1. The bearing journal could move only in the vertical direction (according to the direction of the force). The system described has been shown in Fig. 3.51.

Figure 3.51 also presents the method of discretization of the model. After optimization of the division into finite elements, the model consisted of a total of 9,778 degrees of freedom. For its discretization 2D finite elements marked CPE4RH were used. These were four-node elements with linear shape functions, using reduced integration. Elements of this type are dedicated to the analysis of deformations. The choice of specific finite element and the adopted method of discretization of the bearing were preceded by extensive studies. On this basis, in the presented analyses, optimal parameters of the FEM mesh were adopted, which allowed to obtain accurate results in the shortest analysis time possible. During the studies, deformations of the journal and sleeve were included, as they were treated as rigid elements. The properties of structural materials adopted during calculations are provided in Table 3.3.

As part of the verification of the numerical model, the results of the computer simulation have been compared with the results of experimental studies. The verification procedure proceeded in two stages. In the first stage of verification, the stiffness characteristics of the system obtained during static load were compared. The team conducting the experimental studies noticed that the size of nominal clearance had a large impact on the stiffness of the foil bearing [17]. In light of this, the simulation tests, like the experiment, were performed for bearings with three different values of nominal clearance. The appropriate clearance value was achieved by changing the diameter of the journal, which was successively 38.07, 38.10 and 38.13 mm. Thanks to these changes, the values yielded for the radial nominal clearance were: 0.05, 0.035 and 0.02 mm.

However, it should be noted, that the value of the radial nominal clearance in the foil bearing is a result primarily of the design assumptions. The dimensions of an actual bearing, due to operational inaccuracies, differ quite substantially from the design assumptions. This is mainly due to the manufacturing technology of top and bump foils, which does not allow to make these elements with an accuracy of

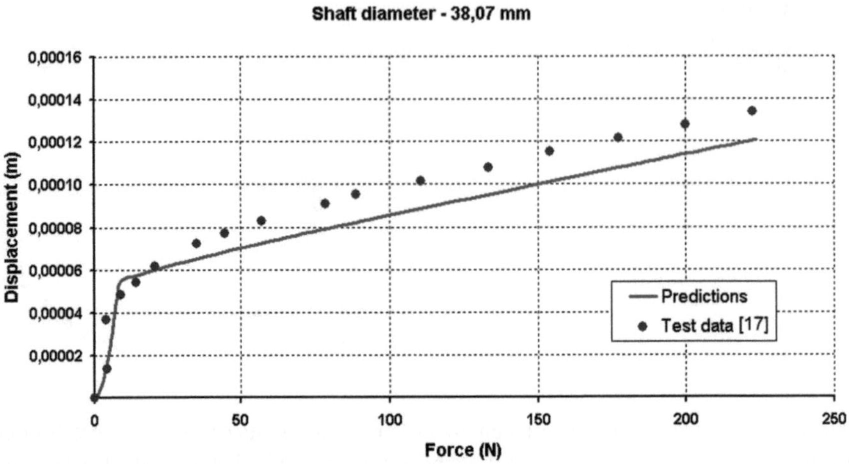

Fig. 3.52 Comparing the experimental and simulation displacements of the foil bearing journal caused by static force (nominal clearance 0.050 mm)

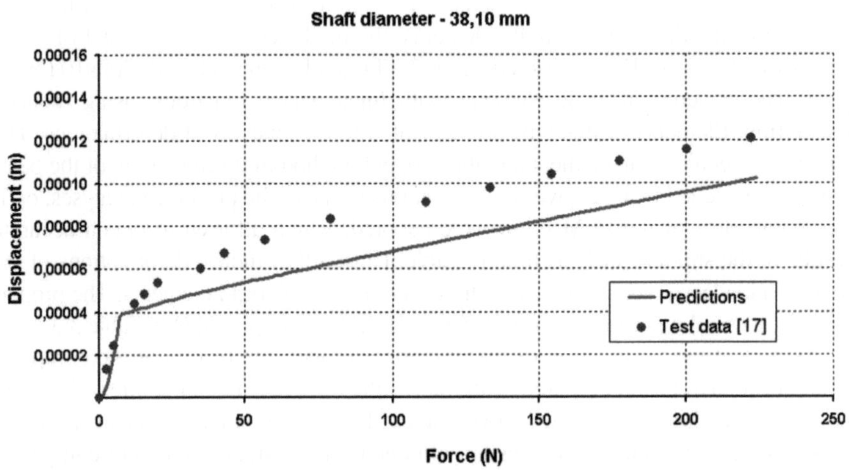

Fig. 3.53 Comparing the experimental and simulation displacements of the foil bearing journal caused by static force (nominal clearance 0.035 mm)

a few μm. Foils which make up the structural bearing layer are elastic elements, and their assembly takes place usually at a certain initial tension. All of this causes that obtaining an assumed dimensional accuracy on the entire perimeter and length of the bearing is practically impossible. The nominal clearance of foil bearings provided in scientific works should, therefore, be regarded as an example value.

The following drawings (Figs. 3.52, 3.53 and 3.54) show a comparison of the results of a computer simulation with the results of experimental studies presented in the work [17]. A characteristic feature of the tested system was the low

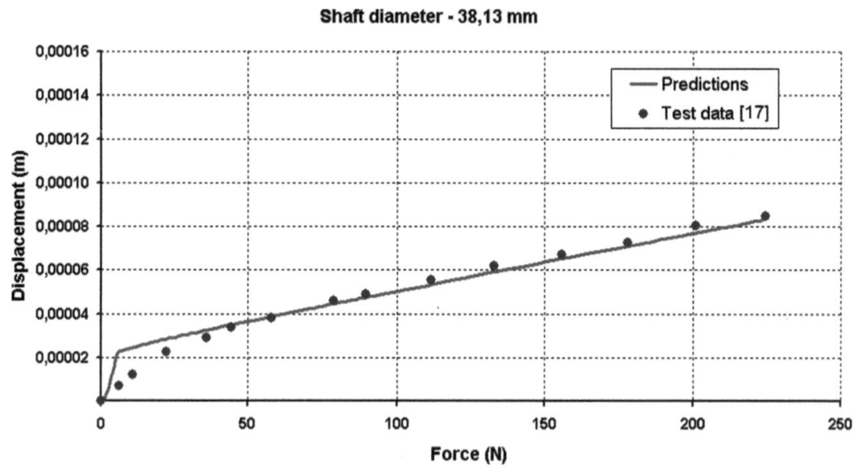

Fig. 3.54 Comparing the experimental and simulation displacements of the foil bearing journal caused by static force (nominal clearance 0.020 mm)

initial stiffness, which stemmed from the presence of a certain clearance. After cancelling the clearance, the tested system increased its stiffness, and its characteristics in the tested load range was similar to the linear one. The comments above concerned three bearings of different journal diameters. In the case of a bearing with a journal with the largest diameter, no clear presence of clearance was observed in the results of the experiment, which confirms the previously described difficulties with obtaining high dimensional accuracy of the top and bump foil.

Compatibility of the results of simulation and the experiment were considered separately for each of the tested bearings. The best compatibility of results was achieved for the system with the smallest clearance. In the load range above 50 N it could be observed that the characteristics, achieved using various methods, overlapped. Some discrepancies occurring at lower loads can be explained by the inaccurate performance of foil assembly, which was the cause of the disappearance of clearance in an actual bearing. In the case of the two other bearings, with a journal of a diameter 38.07 and 38.10 mm, smaller compatibility of characteristics was achieved. In these two cases, the model was a good reflection of the measurement results at a small load in terms of cancelling the radial clearance, which this time also revealed itself at the test rig. At higher load values, there were differences between the calculation results and experiment results of determined displacements, which came up to approximately 20 %. However, it should be noted, that in all the analyzed cases determined by simulation and experimentally, the curves, after cancelling the clearance in the bearing, had a very similar angle of deflection.

In the second stage of static verification, in order to better identify the model of the structural carrying layer of the foil bearing, tests were conducted which

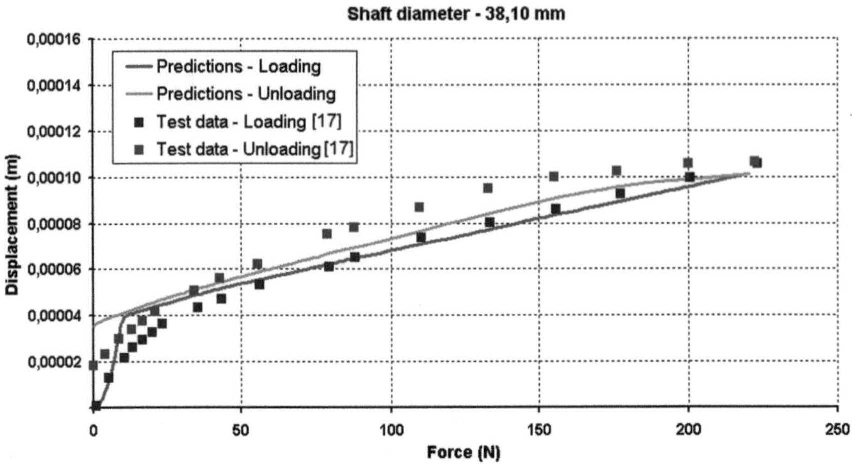

Fig. 3.55 Comparison of experimental and simulation displacements of the centre of the foil bearing journal obtained as a result of the loading and unloading of the system (nominal clearance 0.035 mm)

allowed to assess the model in terms of distributing energy during the loading and unloading process. In the model developed, distributed energy occurred as a result of sliding friction between the cooperating parts of the bearing. For comparing the characteristics of the model with the results of the published studies, in the already cited article [17], in the previously developed model only the procedure of loading the system was modified. During the first second of the analysis the system was loaded linearly with a force of the maximum value 224 N, and in the second of the analysis, this force decreased linearly to 0. The comparison of the obtained characteristics have been presented in Fig. 3.55. The tests discussed were only performed for a bearing with the journal diameter of 38.10 mm.

The characteristics presented in Fig. 3.55 confirms the large reliability of the developed model. Given the complexity of the analyzed system, very good compatibility was obtained of the characteristics during the process of loading the system. Slightly worse fit of characteristics occurred in the unloading of the bearing. The results of the simulation showed that during the unloading, larger values of displacements of the journal corresponded to the same values of force than when loading the system. This was consistent with the results of experimental tests and related to the phenomenon of friction occurring during the displacement of the journal inside the bearing. As a result of the distribution of some of the energy supplied to the system, a so-called hysteresis loop formed. The surface area of the hysteresis loop obtained experimentally was larger than during calculations, which can be explained by the adopted model simplifications, such as omitting internal friction or two-dimensionality [18].

Due to the fact that the tested mechanical system was very complex, and in addition to the deformations of elements of complex geometries, there were also contact phenomena present in it, the results obtained should be considered satisfactory. The test results confirmed the validity of the assumptions adopted when developing the model. Because the developed model can be considered to be reliable, it constitutes a very useful research tool and can be used, among others, for testing the impact of selected parameters on static characteristics of the foil bearing structure.

3.3.3.2 Verification in the Scope of Dynamic Loads

After verifying the correctness of performance of the foil bearing model at a static load, it was analyzed at a dynamic load [19]. Due to the fact that elements of the foil bearing during operation were loaded by variable dynamic forces, it must be taken into account in the model of the bearing. During the verification tests discussed here, a previously characterized foil bearing model was used, which basic dimensions and parameters corresponded to the work [17] and have been provided in Table 3.3. The FEM model of the foil bearing in the variant with dynamic load has been shown in Fig. 3.56.

In order to determine the dynamic characteristics, simulation calculations were planned projecting the experimental tests conducted by Tim et al. [20]. These studies consisted in forcing vibrations in the structural part of the foil bearing using an electromagnetic actuator. During the experiment, the rotor didn't rotate, and its role was played by a shaft fixed rigidly in the lathe spindle. The sleeve of the tested bearing was connected only by a tie, through which harmonic drive from the vibrations actuator was passed. The current value of the driven force was measured using a force sensor, which was placed between the tie and the sleeve. An eddy current sensor was used to measure displacements. The schematic of the testing station has been shown in Fig. 3.57.

The conditions of the experiment were represented during simulation calculations by appropriately provided boundary conditions and the load. The essence of these studies consisted in determining the displacements of the journal in relation

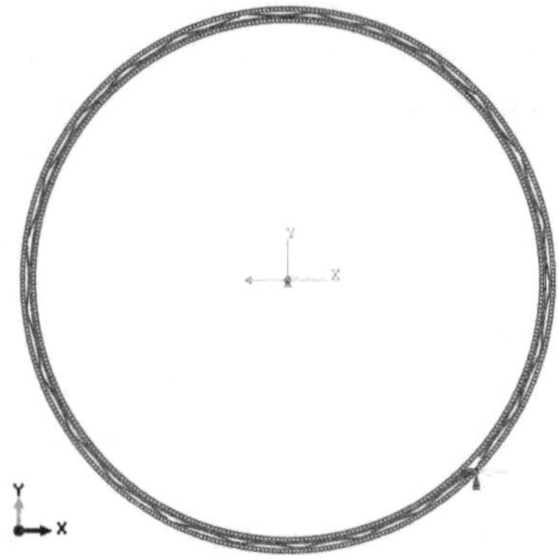

Fig. 3.56 FEM model of foil bearings loaded dynamically

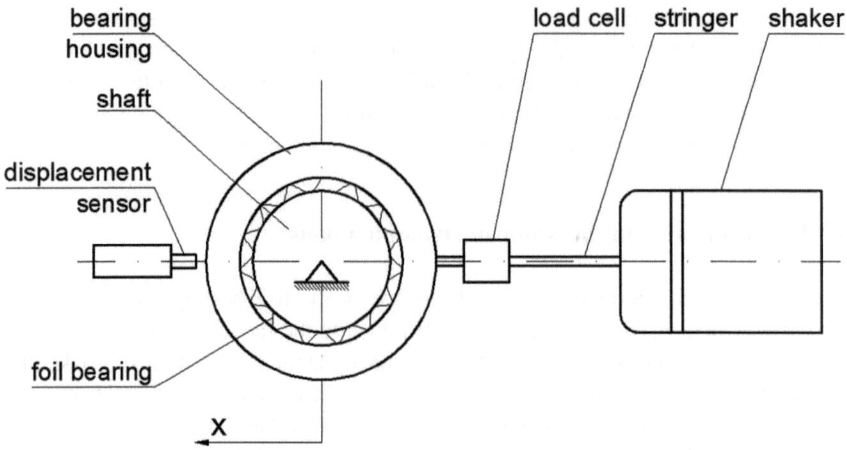

Fig. 3.57 Scheme of test rig to determine the dynamic properties of the foil bearing structure (based on [20])

to the sleeve caused by external load, which were limited mainly by the foil assembly found between the journal and sleeve and the inertia of the bearing's elements. In order to simplify the numerical model, the external load was applied to the journal during analysis (as in actual bearings) and not to the sleeve. And the bearing sleeve was fixed and could not move in any direction. When properly including the mass of the bearing's subassemblies, such an approach makes it possible to obtain results that can be compared with the results of the tests presented in the work [20].

During calculations in ABAQUS, the focus was on the analysis of dynamic responses of the system in the horizontal direction, in accordance with the direction of the drive. The procedure "Dynamic, Implicit" was used with the active option "Nonlinear geometry". The frequency of the drive amounted to 40 Hz, and the amplitude of the driven force 22.5 N (in accordance with the analogous experiment [20]). The value of the force changed according to the sine function. The total duration of the dynamic analysis amounted to 0.2 s, where the harmonic force applied to the bearing active only for 0.1 s (in the range 0.1–0.2 s.) The initial analysis time in the range 0–0.1 s was necessary to determine the contact conditions. The time course of the force loading the bearing has been shown in Fig. 3.58.

In designing the model of the foil bearing structure loaded by dynamic force, it was based on the previously developed and successfully verified static model. As initial calculations have already shown, some parameters of the model required modifications and clarifications. It was also necessary to introduce additional parameters of the model, which in the existing static analysis were insignificant. A key parameter was the mass of the loaded and moving element of the bearing. The mass of additional elements was also important, such as: force sensor, tie, and even thermocouples (used during the experiment for measuring the temperature of part of the bearing). It turned out also that the fairly stiff wires, which were used

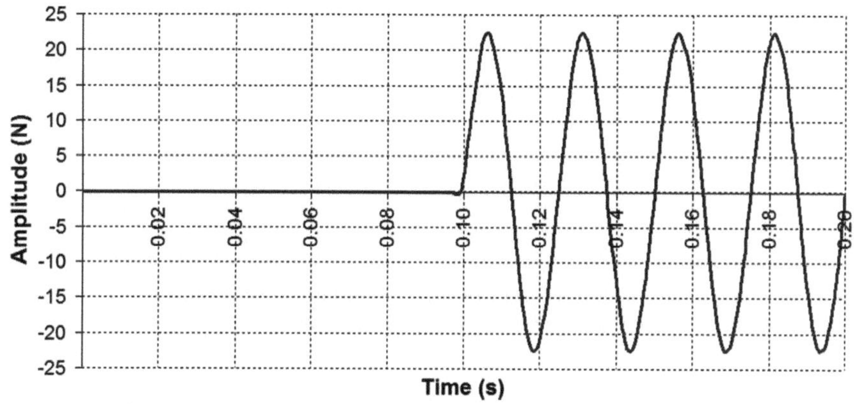

Fig. 3.58 Time characteristics of dynamic load

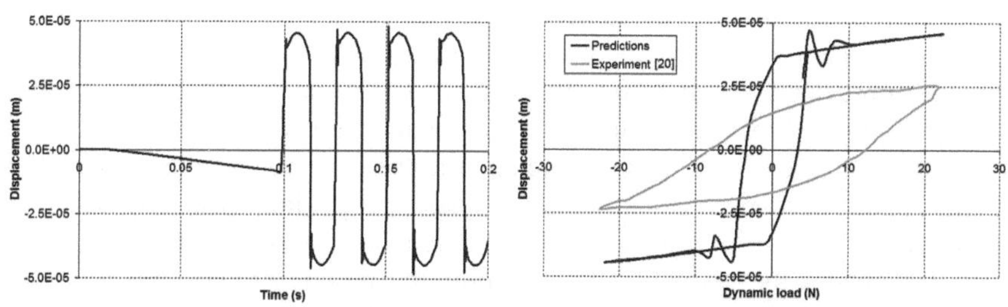

Fig. 3.59 The courses of displacements of the foil bearing journal loaded by dynamic force overtime and depending on the value of the force (results for the model before tuning)

to connect the thermocouples and force sensor, also affected the elasto-damping properties of the system and this needed to be included in the model. The values of these additional parameters, the values of which were not initially known, were chosen in the process of verifying and tuning the model.

Initially obtained characteristics in the form of courses of displacement of the journal during analysis and displacements depending on the value of the load have been presented in Fig. 3.59. It appears from the presented characteristics that the journal, after applying load, first bounced off the foil assembly, and then under the influence of the growing force was again pressed into it. In order to compare the obtained characteristics with the results of the experiment described in the work [20], it was necessary to draw up a graph in the system: dynamic load—displacement. Courses determined by simulation differed significantly from the experimental curves. At a greater load (above 10 N) the displacement of the journal depended only on the value of the load, regardless of whether the system was in the loading or unloading phase. In an actual bearing, a clear difference was observed in the displacements depending on the load phase, which meant the

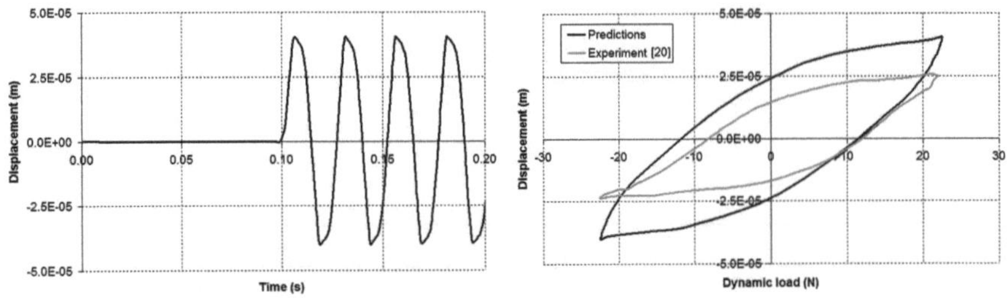

Fig. 3.60 The courses of displacements of the foil bearing journal loaded by dynamic force overtime and depending on the value of the force (results for the tuned model)

ability of the bearing structure to partially distribute energy. It indicated the need to refine the numerical model, so that its characteristics were similar to the actual system.

Modifications to an initially checked model were begun from the identification of the most important differences between the model and test rig. The first potential cause could be neglecting the mass of bearing elements in the model, which in the case of previously conducted static analyses did not matter. Therefore, an additional concentrated mass was introduced into the model, which was connected with the journal's centre of gravity, but in relation to the experiment reflected the mass of the sleeves, sensors, ties and thermocouples, along with wiring. By using the description of the test rig and the result of the series of tests, the value of extra mass was established at the level of 0.5 kg. However, this modification of the system still did not allow to obtain the characteristics similar to the experimental ones. It turned out to be necessary to introduce to the model further modifications consisting in applying additional elasto-damping elements, which were connected with the centre of the journal. After many attempts, the optimal values of damping and stiffness were determined at the level $D = 1{,}000$ N s/m and $K = 4 \times 10^5$ N/m. The course of displacements of the journal over time and depending on the working load, after introducing the described modifications, has been presented in Fig. 3.60. Thanks to the introduction of an additional rigid-damped element in the model, the impact of external elements fixed to the vibrating sleeve was taken into account.

Figure 3.60 enables to directly assess the reliability of the developed numerical model. A direct comparison of the characteristics showed that as a result of tuning the model very similar courses were obtained. The displacements obtained based on the model achieved higher values for the same loads, but the shape of the obtained loops were very close throughout the entire range of work of the bearing. Due to the typical for foil bearings low operational accuracy and the small repeatability of experimental tests caused by temperature changes and surface quality, further attempts aiming to better tune the developed model were not made. The cause of the observed discrepancies was probably the method of taking into account radial clearance in the bearing, which in an actual construction does not exist due to the initial clamp of the journal. Attempts to better tune the model could be made after achieving more results of experimental tests conducted in different conditions.

To sum up the stage of works on the numerical model of the foil bearing discussed in this part of the chapter, it can be concluded that the developed model successfully passed verification and can be used for further analysis. The subsequent subsections of this chapter shall discuss the integration of the structural bearing layer model with the model of the fluid-flow bearing layer. The concept of the comprehensive model of the bearing node will be based on taking into account mutual interactions between the fluid-flow and structure.

3.3.4 Modelling the Fluid-Structure Interaction

Although in literature on the subject reports on developing models of foil bearings can be found, which enable their analysis with a high degree of credibility [21], at IMP PAN developments of their own model of such systems were made. This was primarily because it was necessary to adjust the theoretical model of foil bearings to the specifics of studies carried out within the framework of ongoing study projects. It was assumed that the lubricant medium of developed bearings will be a selected low-boiling medium, not only in gas form but also liquid. Due to the lack of compressibility of the liquid, it was necessary to use a different theoretical description of the phenomena taking place in the oil clearance. All the models discussed in literature concern bearings lubricated by gas, so a medium of high compressibility. Another very important feature that distinguishes the studies conducted from the tests of foil bearings discussed in literature, was the necessity to conduct a very complex designing program and test of the structural bearing layer of the bearing. To analyze such issues specially developed numerical models were needed, created on the basis of the previously discussed, advanced and experimentally verified structural model. To develop a complete model of the entire bearing node, it was necessary also to develop a model of the fluid-flow bearing layer of the foil bearing, as well as its integration with the structural model.

In the developed model of the structural bearing layer of the foil bearing, the full geometry of sensitive elements was taken into account, both of the top and bump foil, and contact phenomena between the cooperating parts, also not omitting friction. The basic result size from the numerical calculations are deformations of the elements of the structural bearing layer of the foil bearing as a result of the effect of static and dynamic loads. Based on these deformations the irregular geometry of the oil clearance is determined. The structural model also allows to conduct strength and thermal analyses.

The analysis of the phenomena occurring in the fluid-flow layer of the foil bearing, depending on the variant of the bearing, is performed by one of two developed methods. At the Institute of Fluid-Flow Machinery of PAN, two alternative models were developed of the fluid-flow of lubricant medium in foil bearings, enabling the analysis of bearings lubricated by a liquid or gas lubricant medium. The model of liquid-lubricated bearings [22, 23] uses the most advanced calculation algorithms of the program NLDW [6], which allows the analysis of the rotor-bearings system in nonlinear dynamics, at the unstable operation of bearings.

In the case of gas foil bearings, at IMP PAN it was decided to develop a mode, which describes the system in the scope of so-called linear dynamics, which has been more widely discussed in chapter one. This approach imposes certain restrictions on the developed model, which causes that it is able to properly reproduce the properties of bearing nodes only in the scope of stable operation of the rotor-bearings system, when so-called "small vibrations" occur around the point of static balance. The trajectory of these vibrations can take on the shape of an ellipse at most. This approach resulted mainly from the fact that the team of IMP PAN did not previously deal with modelling gas bearings, and developing a fully nonlinear model of gas bearings is a task requiring many years of numerical and experimental studies. The model of gas flow of the lubricant medium was developed in cooperation with the team from the Institute of Turbomachinery of the Technical University of Lodz, which has more experience in this subject [24, 25].

Regardless of the state of concentration of the lubricant medium, during the flow analysis in the lubricant space of foil bearings in the first stage, a kinetostatic analysis of the studied system is performed in order to determine the basic parameters of the operation of the bearing node (static load, need for lubricant medium, the position of the journal, pressure distribution in the lubricant space). Then, dynamic calculations are performed, during which in each time step, a new position of the bearing journal is determined, new coefficients of stiffness and damping of the lubricant film, as well as the temporary distribution of pressure in the lubricant space of the bearing.

In order to determine the dynamic characteristics of the whole foil bearing it is necessary to take into account the structural and fluid-flow characteristics of the bearing layer. Because during performance of the bearing (due to the nonlinearity of characteristics and changing load conditions of the system) the basic parameters are constantly changing, during the calculation process, it was necessary to constantly exchange information between modules for structural and fluid-flow calculations. Deformations of the foil assembly caused by an indicated distribution of hydrodynamic or aerodynamic pressure affect the change of the geometry of the oil clearance. The change of the geometry of the oil clearance implicates the change of the distribution of pressure in the fluid-flow bearing layer of the bearing. For these reasons, the calculation process must be done iteratively, and the number of iterations depends on the stability of the performance of the system, convergences of calculation results, as well as assumed accuracy of the analysis. These are very complex issues, requiring for each of the cases detailed analysis and clarification of the precise conditions. A general scheme of the exchange of data between the two programs during the analysis of foil bearings has been shown in Fig. 3.61. Data exchange in the developed model takes place automatically, with the help of a specially developed computer program.

On the basis of the indicated method of characteristics of the bearing, the analysis of the entire rotor system is conducted. To model the rotor at IMP PAN a beam finite element is used (so-called Timoshenko beam), which for simple rotors allows to obtain very accurate results of calculations. Such a model in programs from the MESWIR series is integrated with the model of bearings, which allows to carry out a coherent analysis of the entire system (Fig. 3.62).

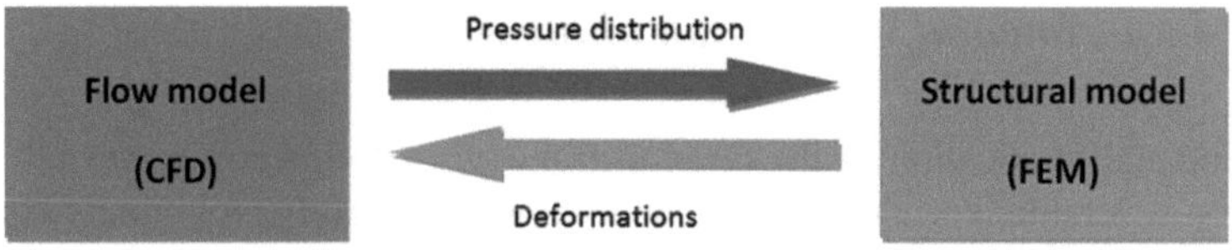

Fig. 3.61 The concept of cooperation of the bearing program with the commercial FEM package during simulation analysis of foil bearings

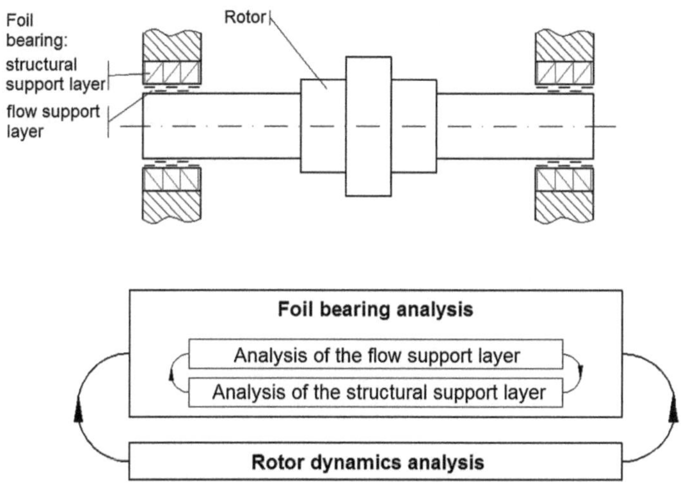

Fig. 3.62 The general calculation scheme proposed for determining the dynamic characteristics of a rotor machine with foil bearings

In accordance with the previously provided information, when modelling the rotor with foil bearings, the concept of modelling based on using proprietary and commercial calculation programs was used. As a result, it was possible to use the specific advantages of the various applications.

In the case of proprietary calculation codes, which were used for modelling the rotor's kinetostatics and dynamics, as well as the flow of the lubricant medium of bearings, the main advantage was the nonlinear model of hydrodynamic bearings and the ability to make any modifications, which caused a continuous improvement of the model. Additionally, the modifications of these programs made it possible to include deformations of the structural bearing layer of the foil bearing. In the case of the commercial program ABAQUS, which was used to model the deformations of the structural bearing layer of the foil bearing, the main advantage was the ability to use ready-made finite elements describing the complex geometry and options enabling to model deformations which included contact phenomena of a sensitive foil assembly of nonlinear characteristics.

3.3.5 Experimental Verification of the Model

Experimental verification of the developed model of foil bearings was attempted based on several independent experiments. However, already preliminary comparisons of the simulation and experimental results showed that ignorance of even a few, but important parameters of the system led to large discrepancies between the results. In connection with this, the focus was on mapping studies characterized in the works [26], which turned out to be described best. During the construction of the model based on the description contained in this publication, concerning only experimental studies, the least doubts were encountered.

The subject of the tests described in the article [26] was a steel rotor with two discs supported on two foil bearings. The experimental tests were performed on a specially built test rig (Fig. 3.63). The foil bearings used during tests were not previously broken-in, as this rig did not allow to carry out the appropriate procedures of breaking-in. On the presented rig, the axial forces were transferred by a two-directional thrust foil bearing, but this bearing was not the subject of the studies. The maximum safe rotational speed of the rotor amounted to 60,000 rpm. The driving took place through a simple air turbine, the blades of which were cut on one of the discs.

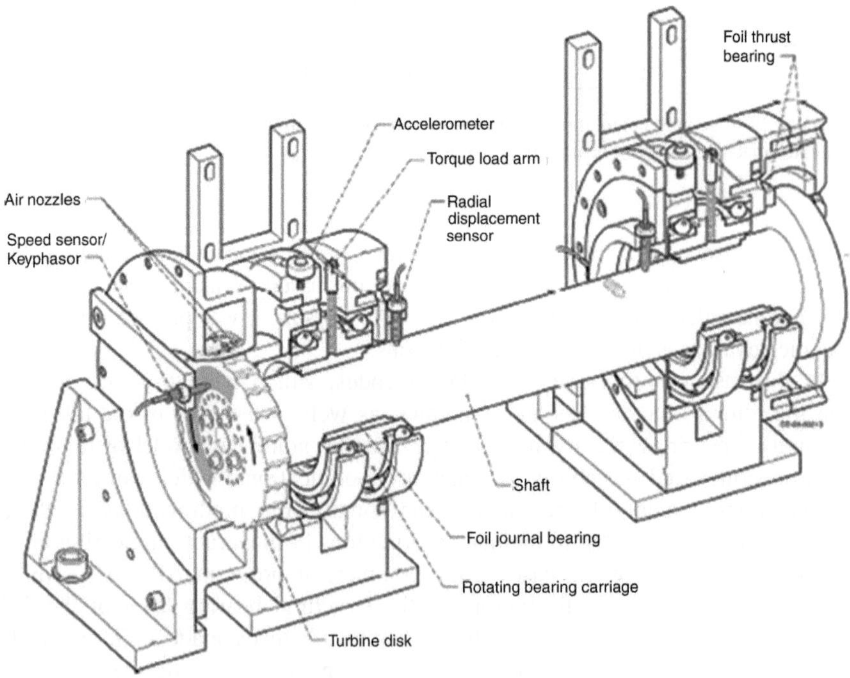

Fig. 3.63 Post for testing foil bearings [26]

Diameter	50.8 mm
Length	40.6 mm
Foil thickness (bump and top)	0.102 mm
Bump height	0.51 mm
Bump pitch	3.17 mm
Bump length	2.54 mm
Bump foil angular extent	360°
Foil material	Inconel 718
Structural loss coefficient	0.2
Top foil stiffening factor	4
Static load, each bearing	33.6 N

Fig. 3.64 Basic parameters of the tested foil bearings [26]

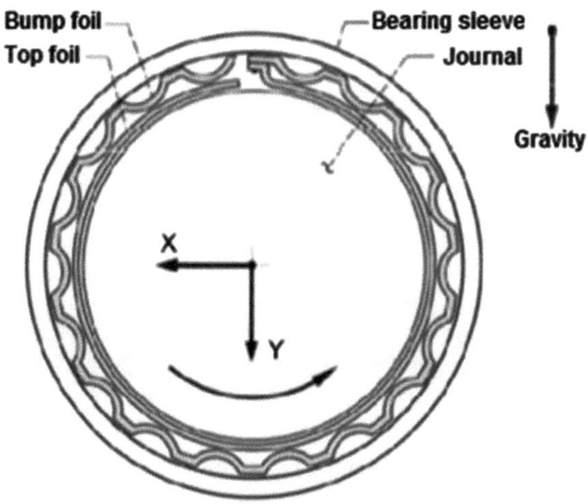

Fig. 3.65 Scheme of the foil bearing of the 1st generation [26]

The basic dimensions of the tested bearings have been provided in a table in Fig. 3.64. The size of the radial clearance was determined experimentally and amounted to 68 μm (right bearing) and 100 μm (left bearing). These differences resulted from the limitations of the technology of making the foils. The geometry of the tested bearings has been schematically presented in Fig. 3.65. These were foil bearings of the 1st generation, which characteristic dimensions of bump foil have been shown in Fig. 3.66.

After specifying the nominal clearance of the foil bearings, tests of the rotor's dynamics were performed. The basic dimensions of the tested rotor have been presented in Fig. 3.67. The weight of the assembled shaft with discs amounted to 67.2 N. Although geometrically this rotor was symmetrical, it should be noted that the driving turbine was on one side, which caused an uneven load on the bearings. In addition, the symmetry of the system was impaired as a result of various radial clearances in the bearings and the different level of imbalance of the discs.

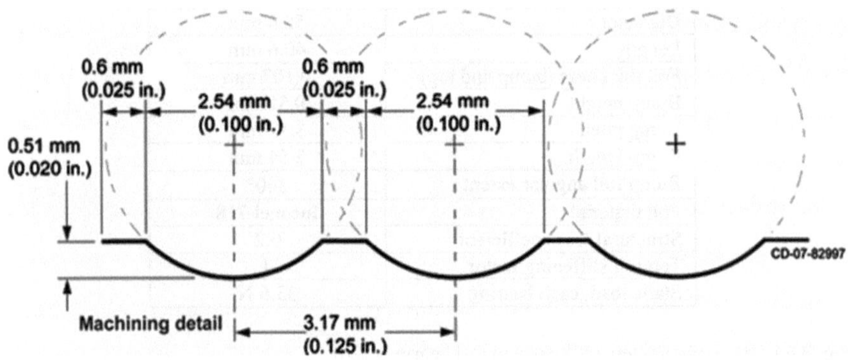

Fig. 3.66 Characteristic dimensions of bump foil of the tested bearing [27]

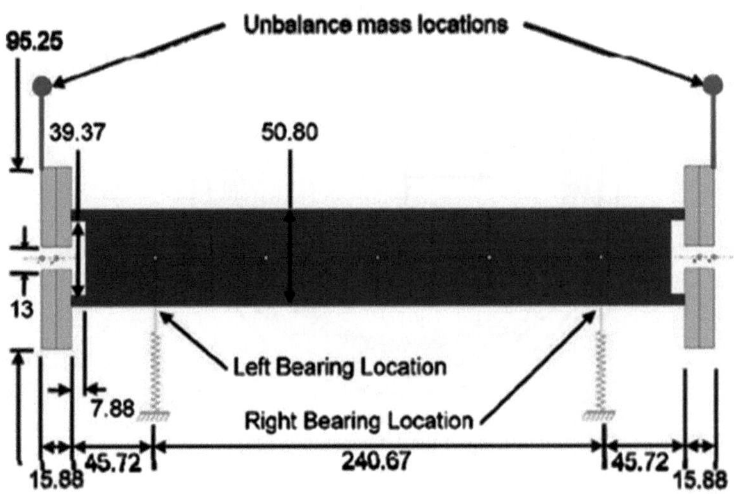

Fig. 3.67 The basic dimensions of the tested rotating system (values provided in mm) [26]

Before performing experimental tests, the rotor was balanced, however, small residual imbalance was placed on both discs (4.2 g mm at 46° and 7.5 g mm at 87°). Imbalance led to obtaining more distinct characteristics of the system and could be entered into the numerical model.

The experimentally obtained characteristics showing the courses of amplitudes of relative vibrations (peak-to-peak), as well as phases for two bearings in the horizontal and vertical direction have been presented in Figs. 3.68 and 3.69. The results of experimental tests have been marked by continuous curves in blue. The relevant characteristics of the model, used by the authors of the studies, were applied on the same graphs in the form of dotted lines in red.

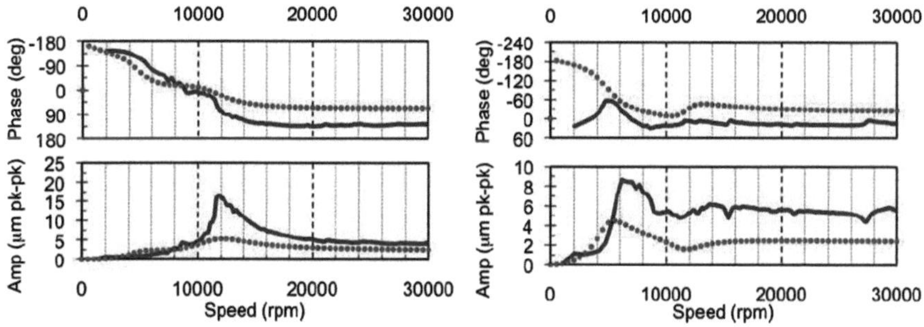

Fig. 3.68 Courses of amplitude and phases of vibrations for the *left* bearing in the *vertical* (left side) and *horizontal* (right side) direction [26]

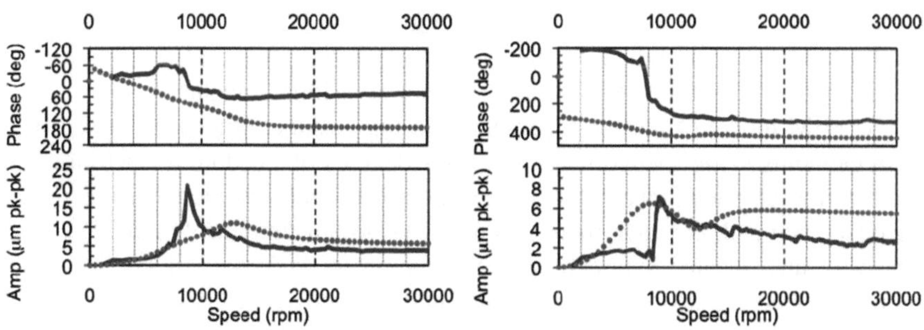

Fig. 3.69 Courses of amplitude and phases of vibrations for the *right* bearing in the *vertical* and *horizontal* direction [26]

The visible maximums on the above courses at speeds close to 10,000 rpm have been explained as critical speeds, which occurred at a slightly different frequency in the horizontal and vertical direction. To comment on the aforementioned courses, certain discrepancies of the simulation and experiment results were noted. The main reasons for the discrepancies were provided as the inaccuracy of balancing the tested rotor, simplified method of modelling imbalance, error in determining clearance in the bearings and inaccurate installation of discs. These suggestions were used when entering data into the numerical model developed at IMP PAN. For the planned verification studies of their own numerical model only experimentally obtained characteristics were used (Figs. 3.68 and 3.69), but the results of the numerical calculations obtained by other researchers could also be valuable material to help explain any differences.

A detailed description of the process of creating numerical models of foil bearings developed at IMP PAN in Gdańsk, has been discussed in the previous sections of this chapter. In light of the above, this subsection provides only the most important information about the model, which was eventually subjected to verification. All the dimensions adopted for the model were consistent with the

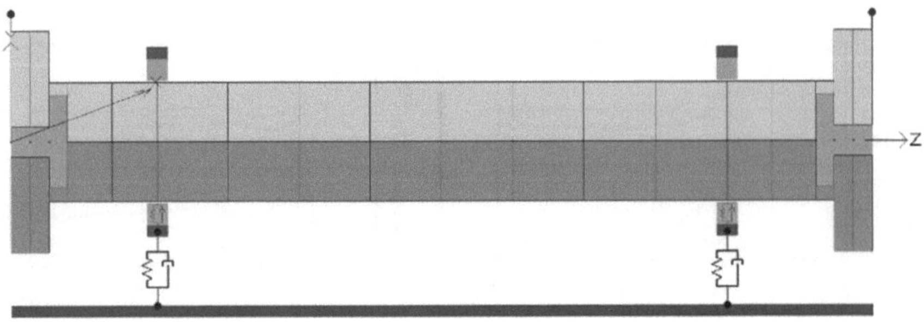

Fig. 3.70 The model of the rotor with foil bearings subjected to experimental verification

dimensions of the rotor in the test rig. The total length of the shaft together with discs amounted to 364 mm, and the shaft diameter was 50.8 mm. The mass of the rotating elements, determined on the basis of dimensions and material density was approx. 6.9 kg. The model also takes into account residual imbalance.

The FEM model of the rotor, presented in Fig. 3.70, consisted of a total of 18 beam elements (Timoshenko beam) with six degrees of freedom at each node. In terms of the geometry, the rotor was symmetrical, and two identical discs were placed at each end, at which there was a hollow with a depth of 24 mm. The bearings were also located symmetrically in relation to the centre of gravity of the rotor. Each bearing was subject to static force directed vertically downwards with a value of 33.8 N, and additionally with forces originating from imbalances. It was assumed that the foil bearings were lubricated by air with the dynamic viscosity of 1.85×10^{-5} Pa s. The dimensions of the bearings were compatible with the actual construction: journal diameter was 50.8 mm, and the width of the bearing 40.6 mm. The size of the radial clearance was chosen in terms of optimal performance of the bearing in a wide range of rotational speeds and amounted to 20 μm. It differed from the clearance determined in the article [26], which amounted to 68 and 100 μm, but in reality certain values of clearances concerned various parameters of the bearing. In the article, in fact, a nominal clearance was determined of a sensitive foil assembly at an initially zero load of the journal. The radial clearance defined for the numerical model related to the initial distance between the journal and sleeve in an operating, loaded gas bearing. The values set out in the article [26] did not have direct application in the analysis of the performance of bearings, as to transfer a load it is necessary to create a fluid-flow bearing layer, which can only arise at an optimal value of the radial clearance.

To properly project the performance of foil bearings, it was necessary to include both the high sensitivity of the structural bearing layer of the bearing, as well as the deformed (as a result of elastic deformations) shape of the oil clearance. The geometry of top and bump foils was developed on the basis of the guidelines referred to in Sect. 3.3.1, and has been presented in Figs. 3.71 and 3.72. Due to the much greater calculating performance (at a similar level of accuracy) a simplified 2D model was used [28]. Essential characteristics of the structure of the foil bearing

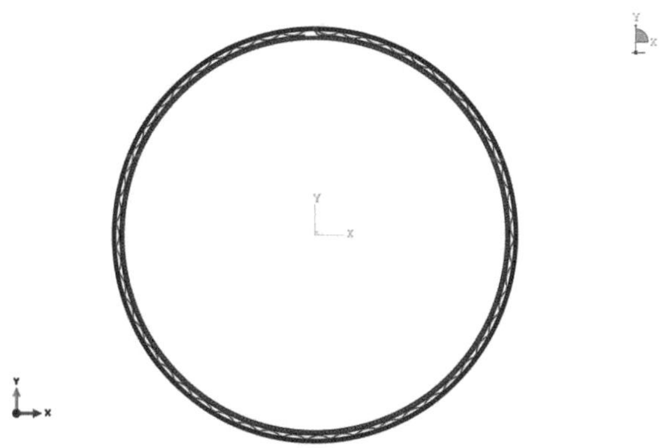

Fig. 3.71 FEM model of the foil bearing structure

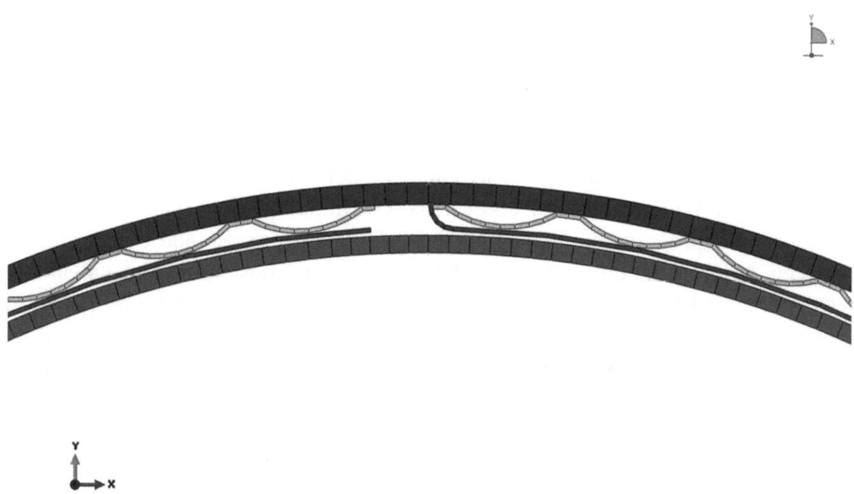

Fig. 3.72 A fragment of the FEM model of the foil bearing structure presented in magnification

were determined using the FEM model, which construction has been discussed in Sect. 3.3.2. By applying the simplification, the number of degrees of freedom of the model amounted to 10,824, at the number of elements being 2,898. Isoparametric finite elements marked CPE4RH [16] were used for discretization. The coefficient of friction for cooperating surfaces was adopted at the level of 0.2.

The determined transverse stiffness of the structural part of the bearing loaded by concentrated force (working vertically downwards) was 6.633 kN/m. This value was determined without including the displacement of the journal in the initial mounting clearance range, which in the developed FEM model is 100 μm. The characteristics of stiffness of the foil bearing structure, determined by simulation, has been presented in Fig. 3.73.

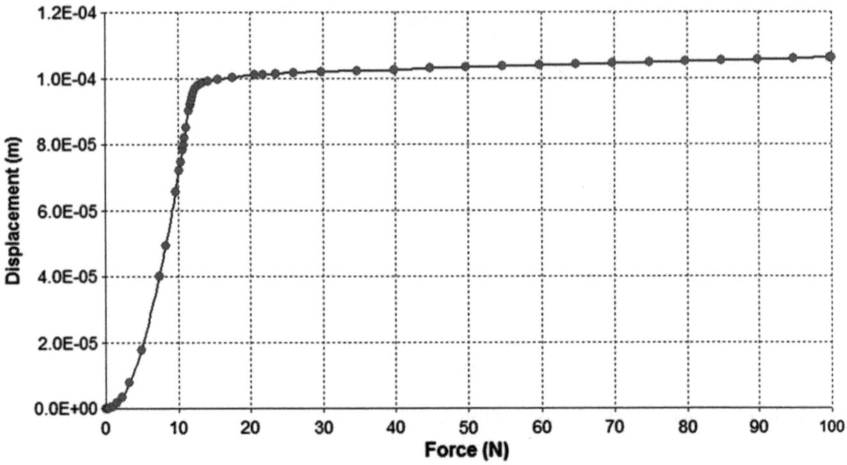

Fig. 3.73 The characteristics of stiffness of the foil bearing structure, determined by calculation, loaded by a concentrated force through the journal (displacement of the journal depending on the working load)

Simulation calculations of the dynamic properties of the tested system using the discussed numerical model was carried out in automatic mode, using a properly prepared program for managing the calculation process and data exchange between different modules. Presented below are the most important calculation results obtained for the model that includes both the elasto-damping properties of the structural carrying layer of the bearing, as well as the deformed shape of the oil clearance. Figures 3.74 and 3.75 show the courses of vibration amplitudes of bearing journals, separately for the horizontal and vertical direction. The results show the vibrations of journals in relation to the sleeves of the bearing, and not in relation to the top foil, which shape was determined by the geometry of the oil clearance.

The courses presented show that the rotor in the tested range of speeds up to 30,000 rpm worked steadily. Critical speed occurred at approx. 9,000 rpm (for the horizontal direction) and at approx. 12,000 rpm (for the vertical direction). At these speeds, the vibration amplitude of the journal (peak-peak) came up to 7 μm in bearing no. 1 and to 16 μm in bearing no. 2, and was greater in the vertical direction. Outside the resonant range, a low level of vibration amplitudes was observed, which indicated the stable performance of the system.

This was also confirmed by the trajectories of the vibrations, which have been presented in Figs. 3.76, 3.77, 3.78 and 3.79 for selected speeds. In these drawings, the vibrations of the journal have been presented against the deformed sleeve, which was limited by the sliding foil. Trajectories of vibrations and deformations have been presented in relation to the radial clearance of the bearing, which during fluid-flow analysis of bearings amounted to 20 μm.

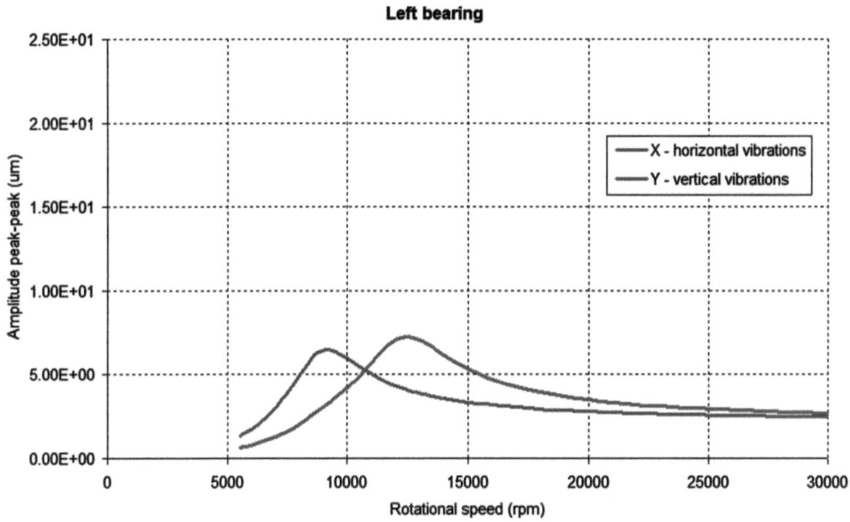

Fig. 3.74 The courses of absolute vibration amplitudes of the journal in bearing no. 1, in the *horizontal* and *vertical* direction

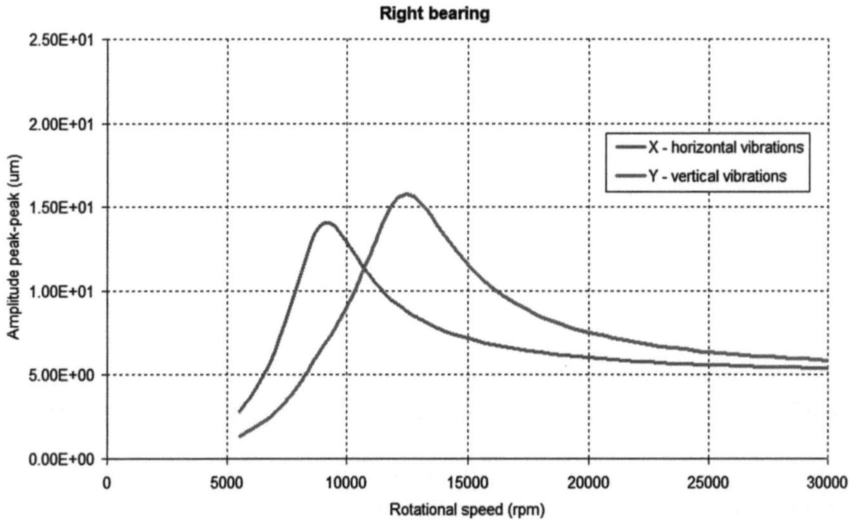

Fig. 3.75 The courses of absolute vibration amplitudes of the journal in bearing no. 2, in the *horizontal* and *vertical* direction

With relatively stiff foils and small deformations, the resulting shape of the sleeve was regular. This enabled to carry out fluid-flow analyses of the medium in the oil clearance of the bearing, which at greater deformations (due to problems

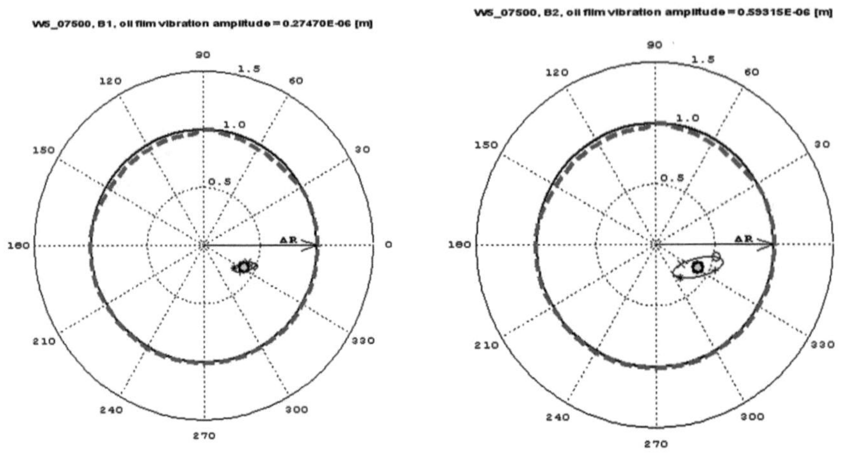

Fig. 3.76 Trajectories of vibrations of journals against a deformed clearance gap at 7,500 rpm

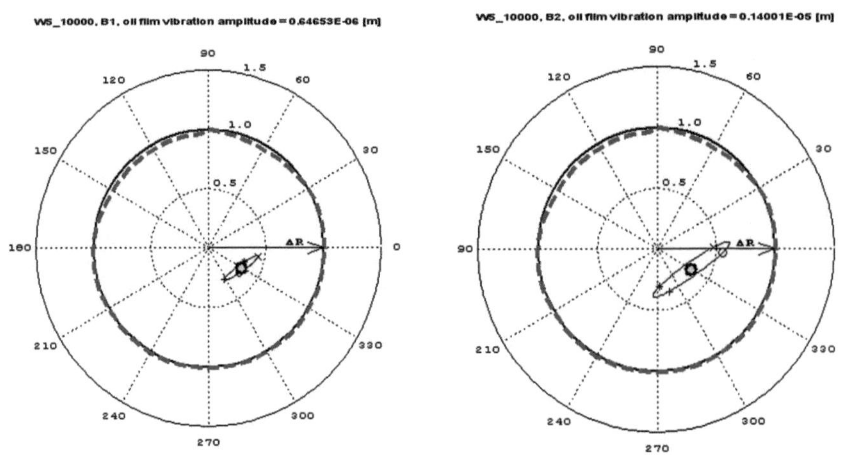

Fig. 3.77 Trajectories of vibrations of journals against a deformed clearance gap at 10,000 rpm

with convergence) were automatically broken. In accordance with the drawings provided above, the pressure acting on the surface of the top foil caused the bending of the foil assembly mainly in the lower part of the bearing. The vacuum prevailing in the upper part of the oil clearance and the load of the lower part of the top foil caused the displacement of the top fragment of the foil in the direction to the centre of the bearing. Generally, deformations included in the analysis of the performance of bearings can be considered as regular.

The results of calculations discussed above were then compared with the results of experimental studies provided in the work [26]. Presented below are the results of comparing the forms of courses of vibration amplitudes (peak-peak), which

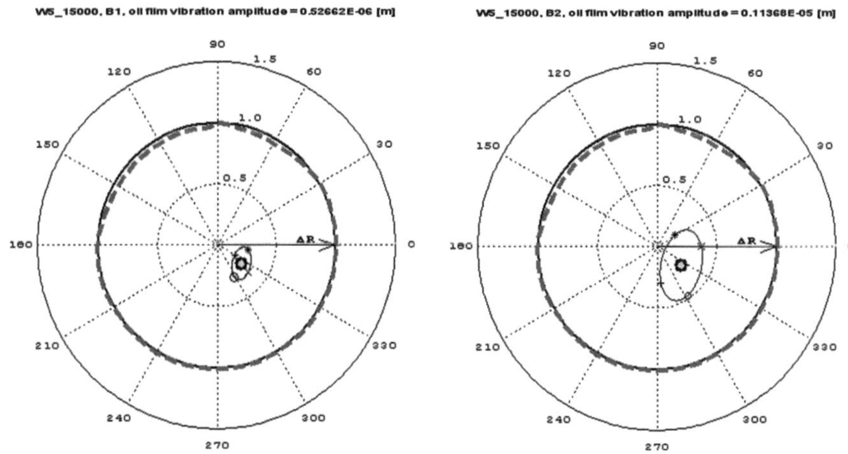

Fig. 3.78 Trajectories of vibrations of journals against a deformed clearance gap at 15,000 rpm

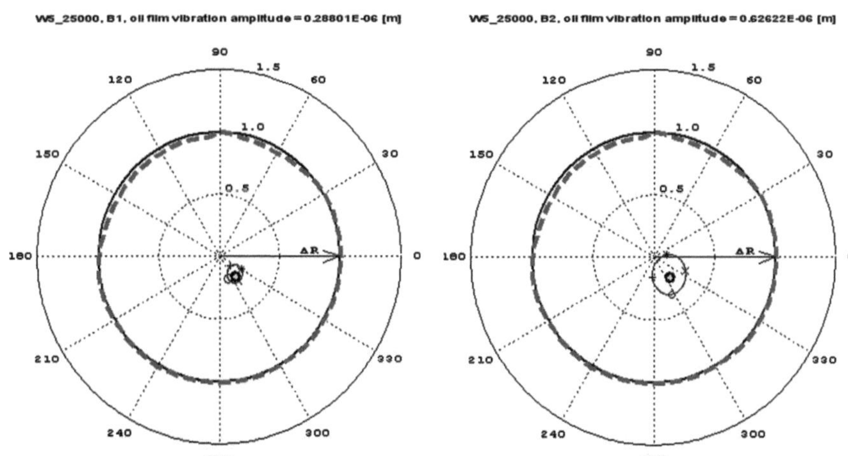

Fig. 3.79 Trajectories of vibrations of journals against a deformed clearance gap at 25,000 rpm

have been drawn up separately for each bearing in two perpendicular directions (Figs. 3.80, 3.81, 3.82 and 3.83). The calculations of gas foil bearings were possible from the speed of 5,500 rpm. Below 5,500 rpm there were not conditions in the bearings for the production of gas lubricant film.

Given the complexity of the analyzed system and the unconventional solution of the test rig, the results of the verification obtained can be considered satisfactory. Above all, the elements of the rig should be noted, which were not projected in the developed numerical model. Firstly, in the test rig, the foil bearings were additionally supported by rolling bearings (to allow lighter start-up and stopping the rotor).

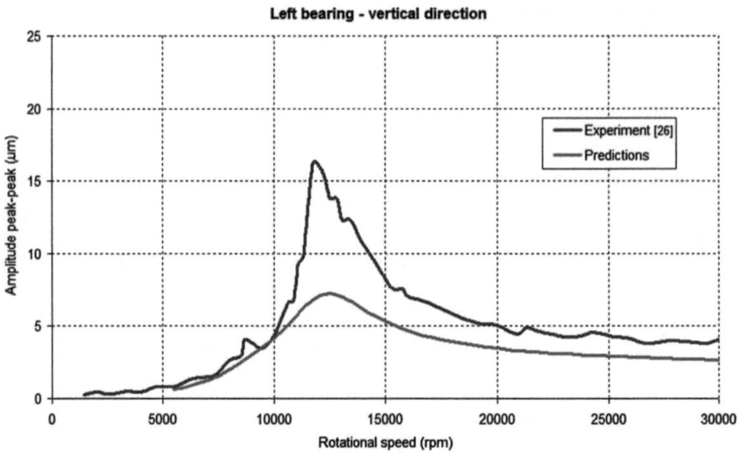

Fig. 3.80 Comparison of the courses of vibration amplitudes of bearing 1 journal in the *vertical* direction, obtained by computer simulation and experimentally

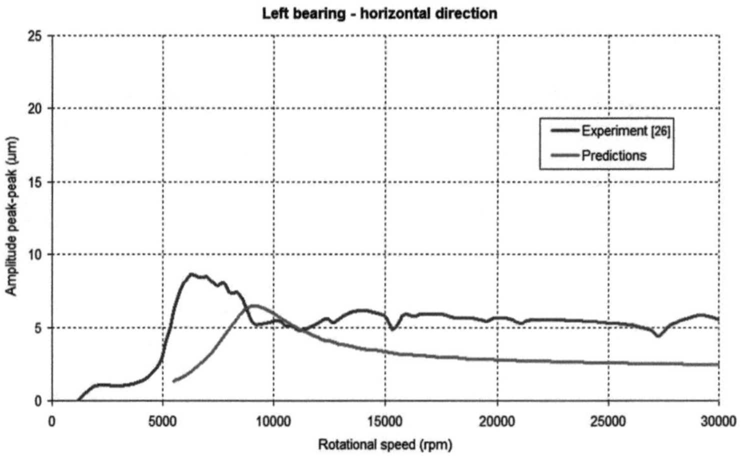

Fig. 3.81 Comparison of the courses of vibration amplitudes of bearing 1 journal in the *horizontal* direction, obtained by computer simulation and experimentally

The properties of rolling bearings were not included in the model, and the entire support was treated as rigid. Meanwhile, rolling bearings, even of the highest quality, are characterized by a nominal clearance and a certain stiffness. Additionally, they changed the distribution of rotating masses and vibration-active elements, and their parts could have been the cause of additional resonances of the test rig and rotor. These factors, due to too many unknowns, were not included in the model. The second important factor affecting the observed discrepancies of simulation and

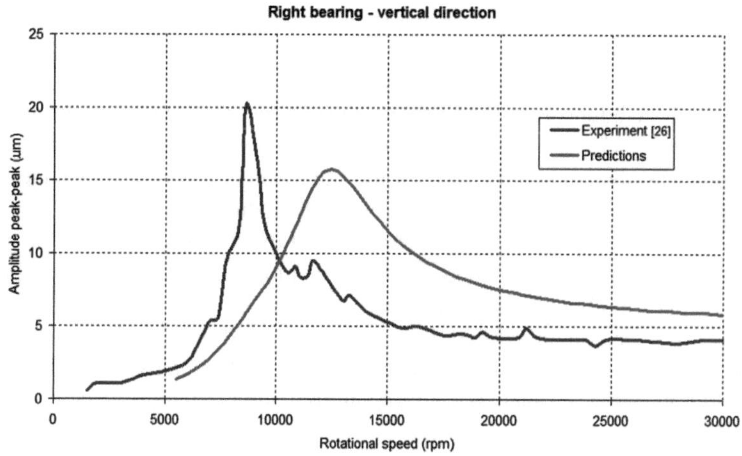

Fig. 3.82 Comparison of the courses of vibration amplitudes of bearing 2 journal in the *vertical* direction, obtained by computer simulation and experimentally

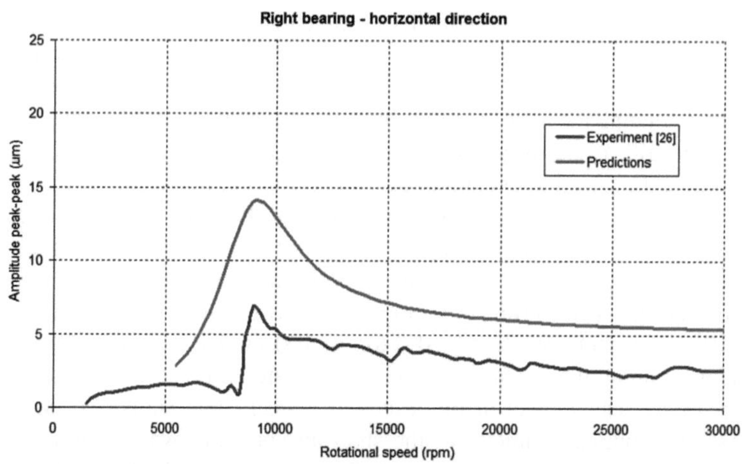

Fig. 3.83 Comparison of the courses of vibration amplitudes of bearing 2 journal in the *horizontal* direction, obtained by computer simulation and experimentally

experimental characteristics was the provisional air turbine, which served to drive the rotor. Notches of the turbine were made on the disc only at one end of the shaft. This caused a disorder in the symmetry of the system and also added more load on one of the bearings. The constant air pressure could also have a stabilizing effect, because it caused an increase in external damping. By comparing the obtained characteristics, the differences in the vibration amplitudes of the systems at higher speeds, in the area of stable operation, should also be noted. The cause of these differences could be a different level of imbalance in an actual rotor, than that declared in the experiment's description, and even its bending. The model also does not include a longitudinal foil bearing.

In spite of the doubts referred to above, the comparisons presented in Figs. 3.80, 3.81, 3.82 and 3.83 show that during the calculations, a similar level of vibration amplitudes was obtained—they were of the same range. It was also managed fairly well to project the critical speeds of the rotor, which during experimental tests were not conclusive, because due to the construction of the test rig, the vibrations moved to another direction and the second support. The model was also unable to generate measurements of characteristic disorders specific for the results. In general, however, it can be concluded that the model of the foil bearing developed at IMP PAN in Gdańsk enabled to obtain reliable and clear characteristics of the system, which did not cause a problem during the interpretation of the results.

3.3.5.1 Analysis of Rotor with Foil Bearings

The advanced model of foil bearings discussed in a previous subsection, integrated with a simple rotor model, was used multiple times to analyze and predict the properties of this type of system [28–30]. This subsection shall present an example of an analysis of a rotor test rig, which was designed and built at IMP PAN. A numerical analysis of the system at the design stage allowed to initially get to know its dynamic properties, including the assessment of the level of vibrations at different speeds and to predict the critical speeds, as well as forms of vibrations of a rotating shaft. It also made possible the selection of appropriate foil bearings of proper stiffness.

The work [29] discusses simulation studies of different variants of the rotor with foil bearings, including the impact of the disc and stiffness of bearings on the characteristics of the system. Here, only a chosen case has been discussed—a rotor with foil bearings of two levels of stiffness. In order to carry out calculations in the MESWIR environment for the tested rotor, a FEM model was created by entering into the program all the geometric dimensions, material properties and parameters of the bearings. The subject of the tests was a steel rotor, supported on two foil bearings. The nominal diameter of the shaft and bearings was 34 mm, the overall length—431 mm. The mass of the shaft determined on the basis of geometric dimensions and density of steel was approx. 3 kg.

The FEM model of the tested rotor (Fig. 3.84) consisted of a total of 20 beam elements with six degrees of freedom at each node. In terms of geometry and division into finite elements, the system was asymmetrical. The end of the shaft on one side had a smaller diameter, suitable for the clutch connecting it with the electro spindle. The lubricant medium adopted in foil bearings was air with a dynamic viscosity of 1.85×10^{-5} Pa s and pressure at the edges equal to atmospheric pressure. The dimensions of both bearings were as follows: journal diameter 34 mm, bearing width 40 mm outer diameter of the sleeve 80 mm, nominal radial clearance 0.02 mm.

Bearings of the 2nd generation were analyzed, as they had several segments of supporting foil. Different levels of stiffness were obtained by modifying the initial

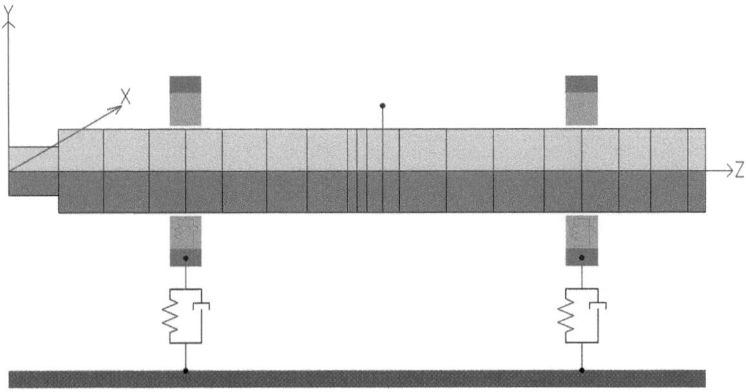

Fig. 3.84 The FEM model of the tested rotor without a disc, supported on two foil bearings

push. The values of stiffness and damping coefficients of bearings adopted in the calculations were determined using the FEM model in ABAQUS [31]. In general it can be concluded that the vertical stiffness of a more sensitive bearing amounted to approx. 10×10^6 N/m, and in the stiff variant it was approx. 100 times greater.

According to the standard ISO1940, the maximum residual unbalance for the analyzed rotor was 15 g mm, but this value turned out to be too small in order to force significant vibration amplitudes of journals during calculations. For the needs of the simulation analysis, unbalance was selected at the level of 51 g mm, which made possible the safe performance of the system, and the largest vibration amplitudes induced in it did not exceed the size of the clearance in the bearings.

For the presentation of the results the program S-MESWIR was used, which helped obtain various types of graphs, such as: relative and absolute vibration amplitudes, trajectories of vibrations of the journal in the bearing, and trajectories of vibrations of the rotor's nodes along the shaft lines. The calculations were carried out for the speed range 9,500–24,000 rpm. All double graphs were presented in an identical manner, which means on the left side are the graphs concerning the left bearing, i.e. the foil bearing mounted nearer the clutch, while on the right side—graphs for the right bearing. On the graphs of courses of vibration amplitudes, blue marks the course of amplitudes for the journal in the left bearing, and red marks the journal in the right bearing.

The following graph (Fig. 3.85) shows the course of vibration amplitudes for the rotor in the case of low stiffness of the foil bearing. In accordance with the following diagram, the vibration amplitudes steadily increased until approx. 22,000 rpm. A larger displacement (peak-peak) of the journal in the sleeve occurred in the right bearing. The maximum value of the displacement occurred for the journal in the bearing at a speed of 22,000 rpm and was approx. 16 μm. For the journal in the left bearing the maximum amplitude value, equal to 14 μm, was achieved at a speed of 21,000 rpm. No results were obtained below 9,500 rpm, which indicated problems with obtaining a stable gas lubricant film in the bearings.

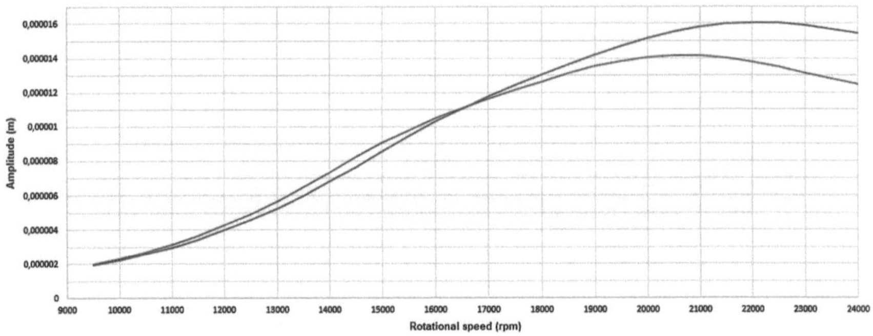

Fig. 3.85 Course of vibration amplitudes (peak-peak) of bearing journals for a rotor with foil bearings of a low stiffness

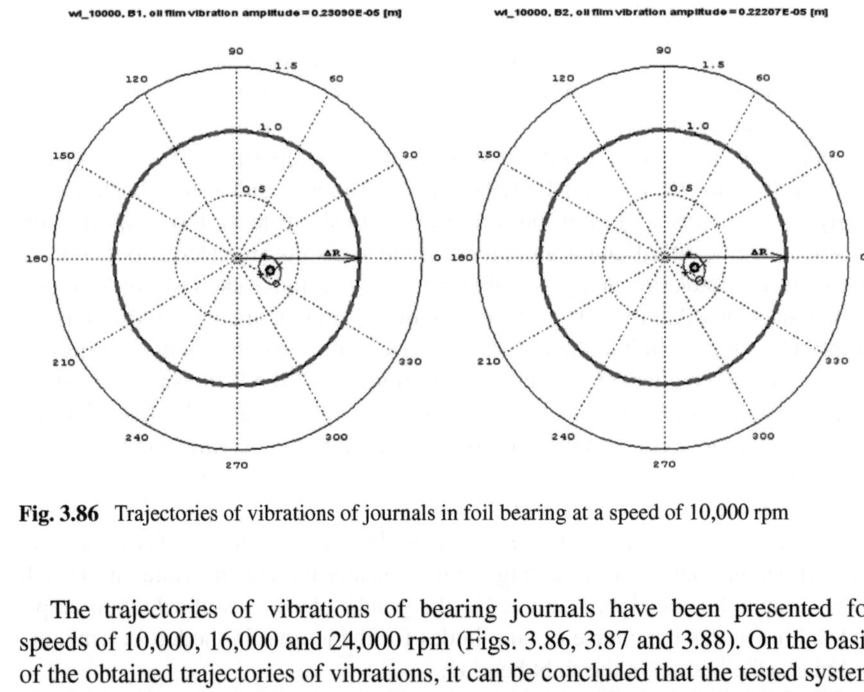

Fig. 3.86 Trajectories of vibrations of journals in foil bearing at a speed of 10,000 rpm

The trajectories of vibrations of bearing journals have been presented for speeds of 10,000, 16,000 and 24,000 rpm (Figs. 3.86, 3.87 and 3.88). On the basis of the obtained trajectories of vibrations, it can be concluded that the tested system worked steadily throughout the entire analyzed range of rotational speeds, but at higher speeds the vibration amplitudes were close to the values of the radial clearance. This was due to the quite large adopted value of unbalance. The trajectories had an elliptical shape and along with the increase in speed their long axis sought to adopt the horizontal direction.

In order to present the nature of the performance of the shaft at a resonant speed, Fig. 3.89 shows the trajectories of vibrations of nodes along the shaft line at a speed of 22,000 rpm. The line of the bent shaft and trajectories of all nodes of the modelled shaft can be observed in the diagram above. The shape of the

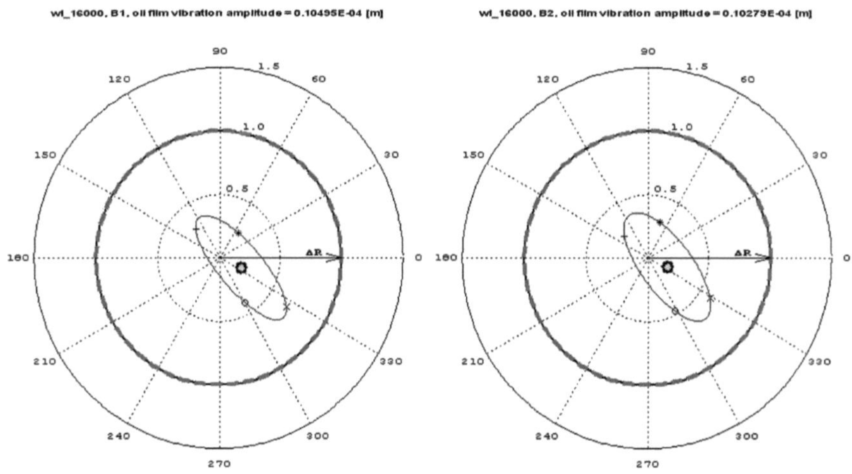

Fig. 3.87 Trajectories of vibrations of journals in foil bearing at a speed of 16,000 rpm

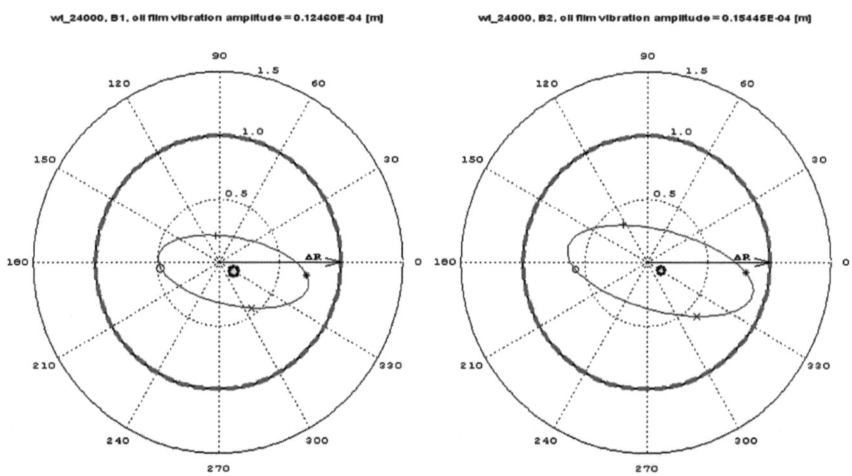

Fig. 3.88 Trajectories of vibrations of journals in foil bearing at a speed of 24,000 rpm

trajectories indicated the so-called translating form of vibrations (cylindrical form) with a slight bending of the shaft.

A similar analysis was conducted for this system with much more stiffer bearings (approx. 100 times). Figure 3.90 shows the course of relative vibration amplitudes of bearing journals at an increased push in the bearings. It can be noticed on the graph that the right end of the shaft achieved slightly higher values of displacements. The maximum values of amplitudes for this variant were 10.8 μm for the journal in the left bearing and 11.5 μm for the journal in the right bearing.

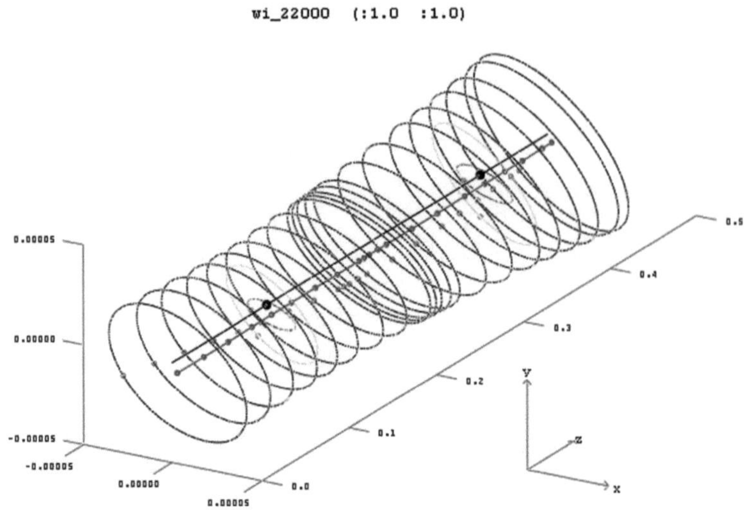

Fig. 3.89 The trajectories of vibrations of rotor nodes along the shaft line at a speed of 22,000 rpm and low stiffness of foil bearing

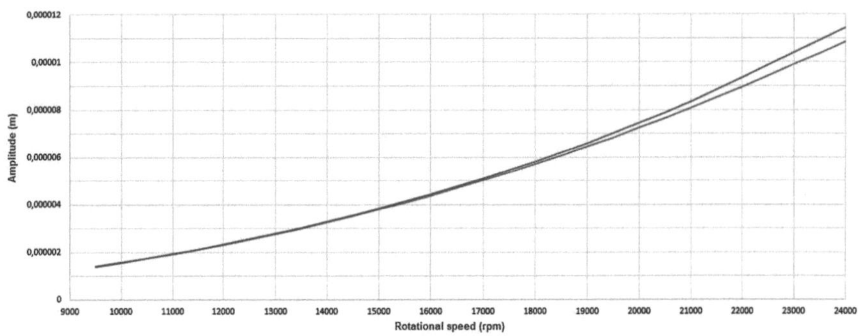

Fig. 3.90 Course of vibration amplitudes (peak-peak) of bearing journals for a rotor with foil bearings with high stiffness

Figures 3.91, 3.92 and 3.93 show the trajectories of vibrations of the shaft's journals at a high stiffness of the foil assembly, at the speeds: 10,000, 16,000 and 24,000 rpm. The system worked steadily throughout the entire speed range. The trajectories of vibrations resembled circles, and grew together with the increase in rotational speed. The system can be described as subcritical, as the resonance did not reveal itself during the tested speed range, and the expected maximum speed of the test rig amounted to 24,000 rpm. The critical speed of the system occurred at a higher rotational speed than the considered maximum speed.

Figure 3.94 shows the form of movement of the rotor at a maximum speed of 24,000 rpm. According to the above drawing, at a high speed the shaft rotated at

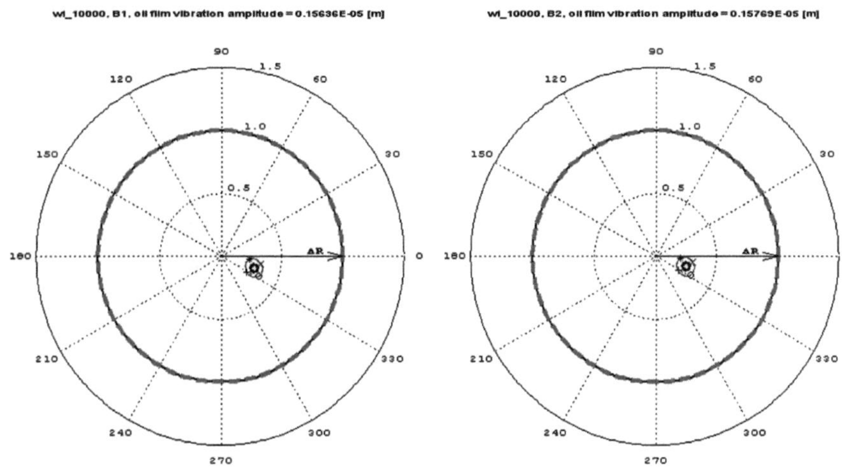

Fig. 3.91 Trajectories of vibrations of journals in foil bearing at a speed of 10,000 rpm

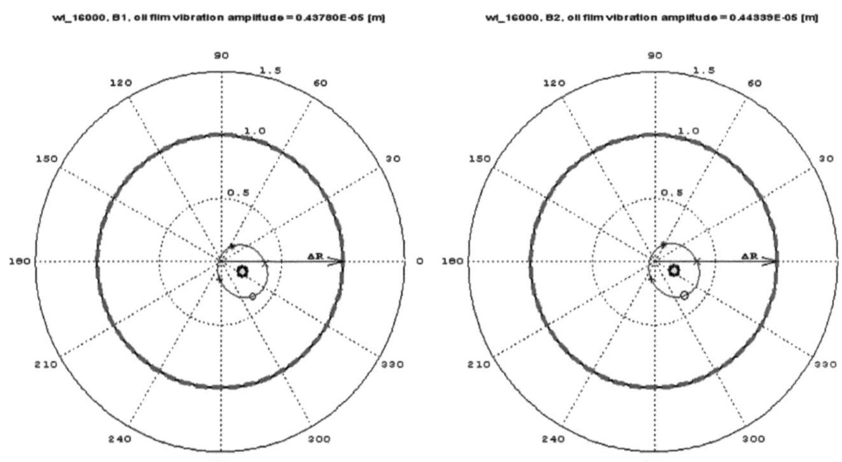

Fig. 3.92 Trajectories of vibrations of journals in foil bearing at a speed of 16,000 rpm

a certain deflection of the centre and free ends. The trajectories of vibrations were regular in shape, which indicated the stable operation of the system.

This subsection presents an example of an analysis of a rotating system with foil bearings made using the numerical models developed at IMP PAN. The created models enable a simulation study of multi-supporting rotors supported on foil bearings of a complex geometry, taking into account the different values of initial clamp. In the discussed case, the model used allowed to assess the impact of the stiffnesses of bearings on the characteristics of the rotor of the test rig. Along with the increase in stiffness of bearings, in accordance with theoretical knowledge, the

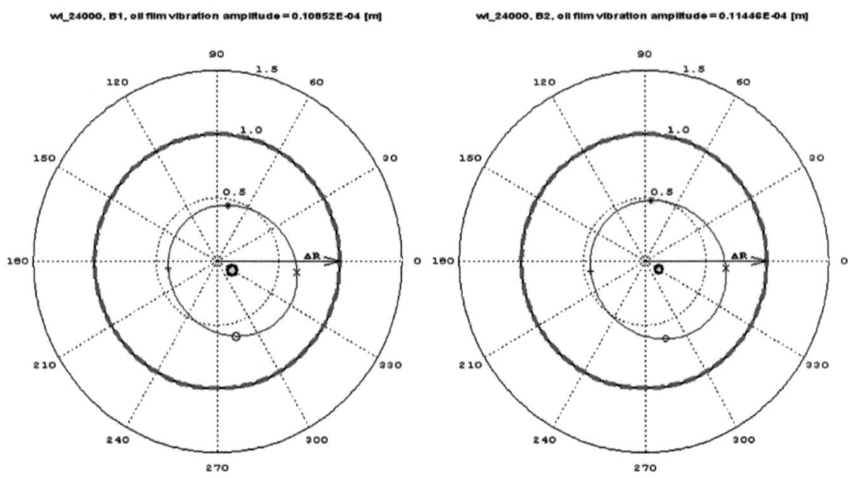

Fig. 3.93 Trajectories of vibrations of journals in foil bearing at a speed of 24,000 rpm

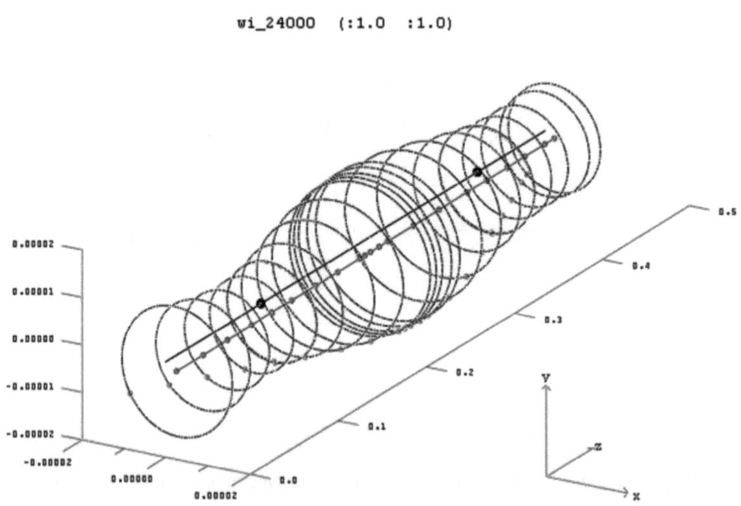

Fig. 3.94 The trajectories of vibrations of rotor nodes along the shaft line at a speed of 24,000 rpm and low stiffness of foil bearing

shifting of resonant speed could be observed in the direction of higher speeds. The vibration amplitudes also decreased. Due to the high stiffness of the shaft and the proper operation of gas foil bearings (after reaching a certain speed) the system in both analyzed variants worked steadily.

As shown above, the numerical models developed at IMP PAN allowed to carry out multi-criterion optimization tests of rotating systems with foil bearings, allowing to predict the properties of such systems by computer simulation.

3.4 Other Bearing Methods

This subsection shall briefly discuss other, modern bearing methods, which can be used, among others, in high-speed microturbines. The directions of development of particular types of bearings will also be discussed, as well as the basic limitations connected with their application. This subsection is mainly to familiarize the reader with the current, very broad possibilities in the scope of bearings, going beyond the widely known and applied solutions.

3.4.1 High-Speed Rolling Bearings

Currently in the world there are more than 150 manufacturers producing high-quality rolling bearings. Each of these companies is constantly expanding the range of their products, while constantly improving them. Figure 3.95 shows an exploded view and typical cross-section of high-speed rolling bearing, with a description of the basic subassemblies and nominal dimensions.

In order to improve the parameters of rolling bearings, their manufacturers are focusing primarily on the following issues:

- construction and material of the basket;
- construction and material of the ring;
- material of rolling elements (e.g. ceramics);
- lubricant medium (consistency, composition).

These treatments will help to increase durability, raise the acceptable operational temperature and the acceptable rotational speed limit.

The primary disadvantage of rolling bearings available on the market is their relatively low, compared with slide bearings, acceptable rotational speed. This is connected with a few problems of which the most important are:

- heat outlet (cooling e.g. only the housing causes a difference in temperatures of the inner and outer ring, thus a different thermal expansion);
- initial stress of the bearing causes (at considerable rotational speeds) the generation of large amounts of heat;
- deformations of rolling elements caused by considerable centrifugal forces cause a change in the clearances in the bearing;
- problems with lubrication (at higher speeds the lubricant medium changes its parameters, which is the result of a strong dependency of viscosity on temperatures and pressures).

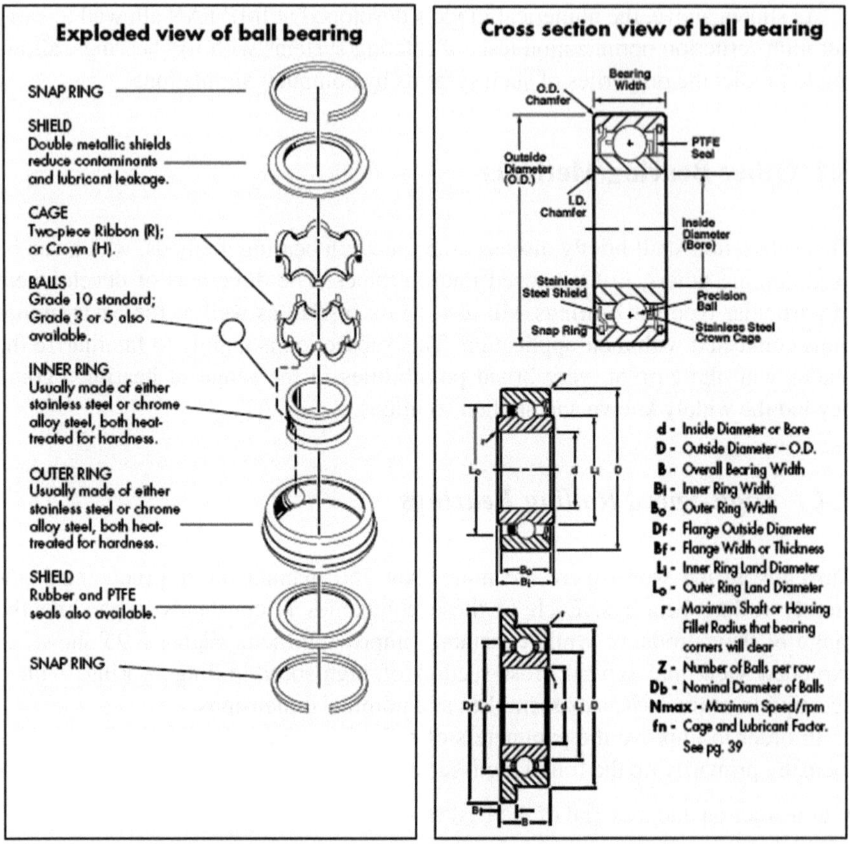

Fig. 3.95 The exploded view and cross-section of example high-speed rolling bearings [32]

Efforts are being undertaken to reduce the typical drawbacks of rolling bearings, which contributes to their broader applications. Discussed below are example solutions proposed by manufacturers of rolling bearings.

The traditional material used for the production of rolling bearings is stainless steel (AISI 440C) and steel with a high content of chromium (SAE 52100). The operating temperature of bearings made from such materials usually does not exceed 180 °C [32]. For example, for extreme applications, the company NHBB suggests hybrid rolling bearings (in which the rings are made of steel, while the rolling elements from silicon nitride) or ceramic bearings [32]. This material, in addition to resistance to higher temperatures, ensures longer life, lower resistance to friction, lower weight and higher stiffness. Thanks to the smaller weight of rolling elements, the centrifugal forces acting on these elements decrease, and during their work, less heat is produced. Figure 3.96 presents a graph which shows the impact of the material used for the production of rolling elements on the operational

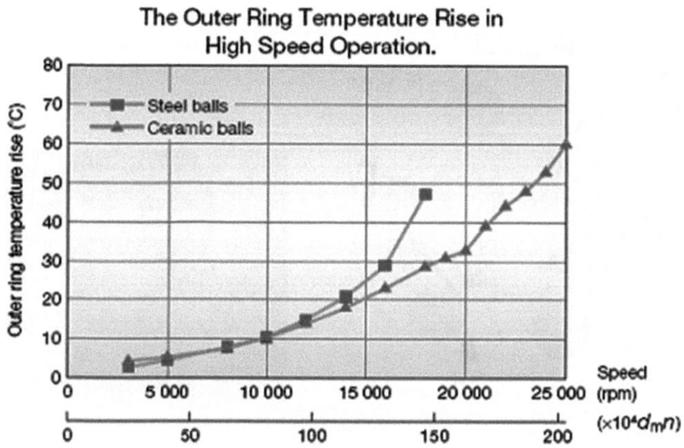

Fig. 3.96 Impact of material of a rolling element on the temperature of the outer ring [32]

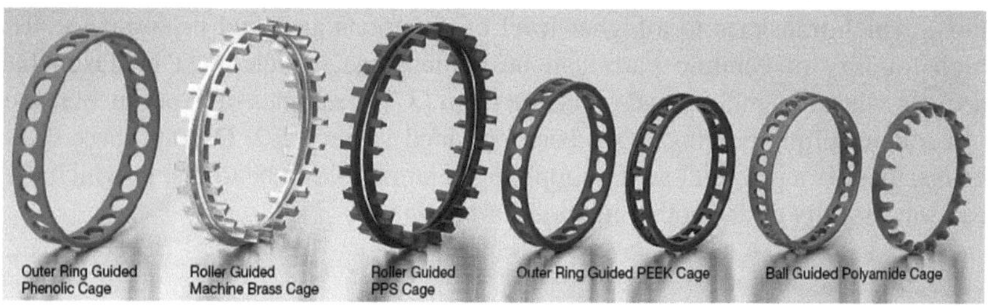

Fig. 3.97 Various baskets determining the location of rolling elements [33]

temperature of the outer ring. Bearings are also produced in which the rolling elements made of steel are coated with other materials, e.g. titanium carbide. Thanks to this, performance of the bearings is possible at higher temperatures and higher stiffness of the bearings.

Companies manufacturing rolling bearings focus greatly also on the construction of baskets determining the location of rolling elements. Figure 3.97 shows examples of solutions offered by the company NSK [33].

In the scope of rolling bearings, there are also solutions proposed that have rolling elements in shapes other than a sphere. The details of such solutions can be found in catalogues of manufacturers [33–35]. There are also some very interesting solutions, in which single rolling elements are replaced by smaller bearings. An example of the concept and implementation of such a design has been shown in Fig. 3.98.

As own studies of ceramic rolling bearings [37] have shown, their most important advantage in relation to microturbines is the ability to work at a higher temperature without the typical lubricant medium. However, experimental studies have shown some disadvantages, which limits their widespread use. The most

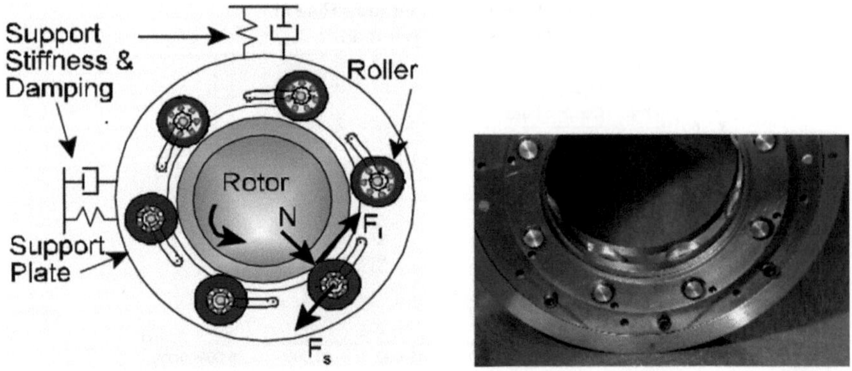

Fig. 3.98 The concept and implementation of an innovative rolling bearing [36]

important of them is the rather large nominal clearance (bigger than in steel bearings), which translates to a higher level of vibrations and loud performance. The high fragility of ceramic parts can be burdensome, which must be taken into account when assembling and replacing them [37]. Examples of experimental tests of ceramic rolling bearings have been provided in Sect. 5.2. Despite these drawbacks, there is a potential area of applying ceramic rolling bearings, in which they can successfully replace other methods of bearings.

3.4.2 Magnetic Bearings

Magnetic bearings are non-contact bearings, in which the phenomenon of magnetic levitation is used to generate carrying force, which consists in the non-contact lifting of bodies in a magnetic field. In active magnetic bearings this force is produced by electromagnets called electromagnetic actuators. Controlled magnetic bearings are mechatronic devices, combining in themselves issues of mechanics, electronics, computer science and automation [38]. The main features distinguishing magnetic bearings compared with rolling bearings and slide bearings are: no friction, no lubricating medium and the ability to operate in any environment. Thanks to these properties, such bearing systems are increasingly used in technology. The following drawing (Fig. 3.99) provides an example of the construction of the magnetic bearing.

Thanks to the ability to control the properties of magnetic bearings, they are used even for the performance of automation tasks. Systems of active control are also used in the bearings of rotor machines of individual operational requirements. The use of such a system allows to [38]:

- increase the efficiency of the device (no friction);
- eliminate the lubricant medium and sealing elements related with it;

Fig. 3.99 Example construction of the magnetic bearing [39]

- the implementation of a closed, airtight construction of the device;
- performance of the device at high rotational speeds;
- performance of the device in a wide range of temperatures;
- performance in different conditions (e.g. vacuum).

Table 3.4 presents a comparison of the most important features of magnetic bearings with classic bearings (slide and rolling).

A unique feature of magnetic bearings is the ability to change their stiffness and damping, which allows to adapt their properties to the current rotational speeds of the rotor. For example, systems for which the first resonant speed is approximately 7,000 rpm, thanks to active control, are able to work at the nominal speed of 10,000 rpm [40]. The following diagram (Fig. 3.100) shows an example of a system supported on magnetic bearings. In this case, a rotor with a mass of approximately 10,000 kg and length of 10 m was set on the magnetic bearings, at a rotational speed of 6,000 rpm. This example is excellent in showing the very great potential of modern magnetic bearings.

The most important disadvantage of magnetic bearings is their complex construction and the need to apply complex control systems with feedback. Therefore, the construction of these bearings involves considerable costs. The following drawing (Fig. 3.101) provides a scheme presenting tasks which the supervision and control system of magnetic bearings can perform.

In accordance with the presented drawing, depending on the complexity of the control system of magnetic bearings, it can perform a variety of tasks, including in the scope of machine diagnostics. They are performed through data acquisition from measuring systems and the modification of control signals, sent to executive elements. The most important task of such a system is always the stabilization of the rotor's position in bearing space, which has been schematically shown in Fig. 3.102.

Table 3.4 Comparison of the most important features of classic bearings and magnetic bearings [38]

Characteristic	Classical bearing	Magnetic bearing
Rotational speed	Limited by critical speed and stability	To 300,000 rpm (it depends on rotor's diameter)
Leading	Theoretically unlimited (growing with rotor's diameter)	Theoretically unlimited (growing with bearing's diameter)
Noise and vibration	Variable (depends on kind of bearing and operating time)	Eliminated
Environment	Limited (temperature, chemical resistance)	Almost unlimited
Reliability	Variable (depends on loading and conditions of operation)	High
Cost	Low	High
Maintenance costs	Usually high	Usually low
Lubrication	Required seals, lubrication systems, constant control	Not present
Possibility of diagnostics	The need for specialized measuring equipment	Continuous monitoring of vibration, current, temperature, etc. due to the measurement and control system
Energy consumption	Variable, depending on operating conditions and type of bearings. Typically the energy-consuming cooling systems and lubrication are required	Saving of up to 90 % of the energy consumed in comparison to the same machine with conventional bearings

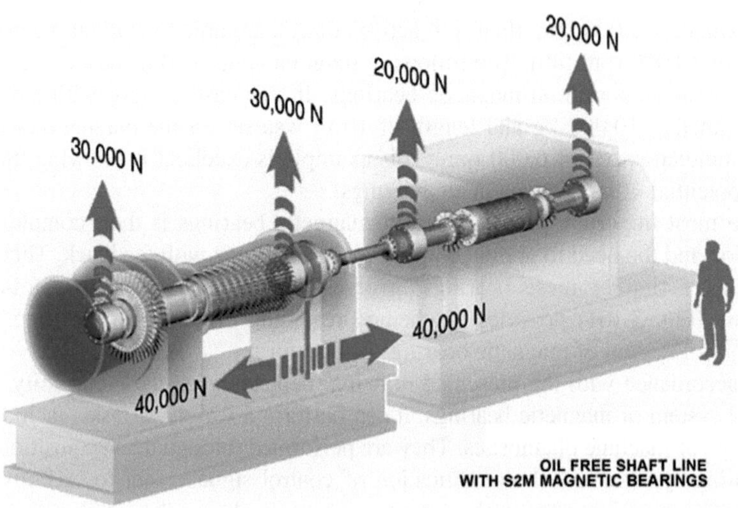

Fig. 3.100 An example of the rotor with magnetic bearings [41]

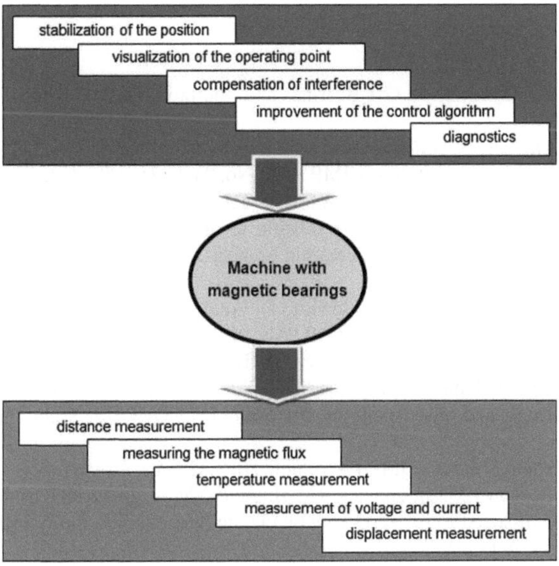

Fig. 3.101 Tasks of the control system of magnetic bearings

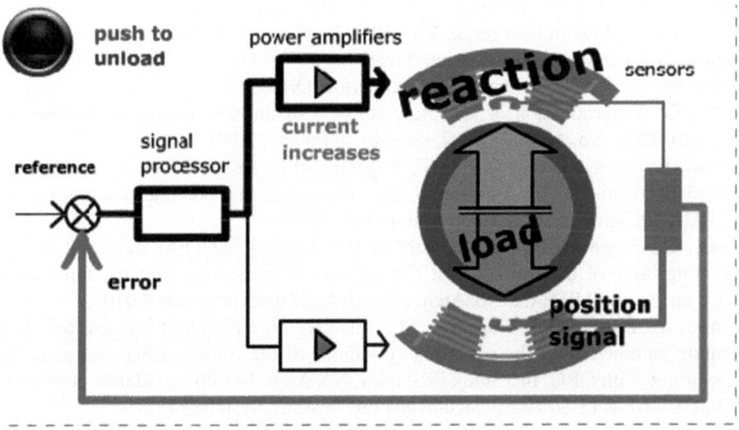

Fig. 3.102 The scheme of performance of the control system of a magnetic bearing in the scope of stabilizing the rotor's position in bearing space [42]

Currently, in the world's scientific and industrial centres, works are in progress on perfecting the construction and control systems of magnetic bearings. The main focus is on simplifying the construction and ensuring stable operation of the rotor at minimum configuration of the control system. Today's state of knowledge allows to use such solutions in systems requiring high precision, reliable work and continuous diagnostics.

References

1. Kaźmierski, K., Krysiński, J.: Gas Bearings and Microturbine Propulsions. WNT, Warsaw, in Polish (1981)
2. Żywica, G.: Analysis of constructional solutions of microbearings available in the literature and company catalogues. Int. Study IMP PAN, No. Arch. 7171/07, Gdańsk, in Polish (2007)
3. Kiciński, J., Żywica, G., Olszewski, A.: The analysis of constructional solutions of bearing systems for turbogenerator's rotor in relation to operating conditions of low-power steam power plant. Part 1: slide, foil and ceramic bearings. Int. Study IMP PAN, No. Arch. 78/09, Gdańsk, in Polish (2009)
4. Miąskowski, W., Nalepa, K., Pietkiewicz, P.: Analysis of construction materials used for making of microbearings. Int. Study IMP PAN, No. Arch. 7679/07, Gdańsk, in Polish (2007)
5. Miąskowski, W., Nalepa, K., Pietkiewicz, P.: Detailed analysis of constructional solutions of microbearings including rolling bearings, slide bearings, lubricated with liquid and gas, magnetic bearings, and other solutions. Int. Study IMP PAN No. Arch. 7306/07, Gdańsk, in Polish (2007)
6. Kiciński, J.: Rotor dynamics. IMP PAN Publication, Gdańsk (2005)
7. Kiciński, J., Rybczyński, J., Żywica, G., Bogulicz, M., Czoska, B.: Simulation studies of small-sized, high-speed rotors supported on classic bearings. Int. Study IMP PAN, No. Arch. 648/08, Gdańsk, in Polish (2008)
8. Kiciński, J., Bogulicz, M., Żywica, G., Czoska, B.: Simulation studies of small-sized, high-speed rotors supported on slide bearings with siphon pockets. Int. Study IMP PAN, No. Arch. 45/09, Gdańsk, in Polish (2009)
9. Hashmat, H., et al.: Low-friction wear-resistant coatings for high-temperature foil bearings. Tribol. Int. **38**, 1059–1075 (2005)
10. Kiciński, J., Żywica, G.: Bearing systems for high-speed microturbine rotors used in CHP installations. Archiwum Energetyki **XXXVIII**(2), 137–145, in Polish (2008)
11. Agrawal, G.L.: Foil Air/Gas Bearing Technology—An Overview. ASME International Gas Turbine and Aeroengine Congress and Exhibition, Orlando (1997)
12. Żywica, G.: Determination of substitute stiffness of the foil bearing supporting layer. Int. Study IMP PAN, No. Arch. 7172/07, Gdańsk, in Polish (2007)
13. Kiciński, J., Żywica, G., Banaszek, S., Bogulicz, M., Czoska, B.: Modelling of foil bearing sleeve deformations using commercial and non-commercial computing software. Int. Study IMP PAN, No. Arch. 222/08, Gdańsk, in Polish (2008)
14. Żywica, G., Bagiński, P.: Development of two-dimensional FEM model of the structural supporting layer of the foil bearing. Comparison of the results with the three-dimensional model. Int. Study IMP PAN, No. Arch. 238/2010, Gdańsk, in Polish (2010)
15. Kiciński, J., Żywica, G., Bogulicz, M.: Analysis of the impact of selected design and operating parameters on the operating conditions of the rotor-bearing system in the steam microturbine with ORC. Int. Study IMP PAN, No. Arch. 133/2012, Gdańsk, in Polish (2012)
16. ABAQUS CAE 6.13 Software documentation, Dassault Systemes (2013)
17. Rubio, D., San Andres, L.: Bump-type foil bearing structural stiffness: experiments and predictions. ASME J. Eng. Gas Turbines Power **128**, 653–660 (2006)
18. Żywica, G.: The static performance analysis of the foil bearing structure. Acta Mech. et Autom. **5**(4), 119–122 (2011)
19. Żywica, G.: The dynamic performance analysis of the foil bearing structure. Acta Mech. et Autom. **7**(1), 58–62 (2013)
20. Kim, T.H., Breedlove, A.W., San Andres, L.: Characterization of a foil bearing structure at increasing temperatures: static load and dynamic force performance. J. Tribol. **131**, 041703 (2009)
21. Kiciński, J., Żywica, G., Bogulicz, M.: Development of assumptions and general concept of modelling of foil bearings and their components. Analysis of the possibilities of using ready-made models of foil bearings. Int. Study IMP PAN, No. Arch. 735/2010, Gdańsk, in Polish (2010)

22. Kiciński, J., Bogulicz, M., Żywica, G., Banaszek, S.: Modification of KINWIR program for kinetostatic analysis of rotor-bearing system, in terms of taking into account changes in oil gap geometry occurring in foil bearings. Int. Study IMP PAN, No. Arch. 370/2011, Gdańsk, in Polish (2011)
23. Kiciński, J., Bogulicz, M., Żywica, G., Czoska, B.: Development of a program for the integration of the calculation of foil bearings taking into account the deformation of the foils caused by hydrodynamic pressure. Int. Study IMP PAN, No. Arch. 436/2011, Gdańsk, in Polish (2011)
24. Tkacz, E.: Numerical model of aerodynamic gas bearing—description of the GAZBEAR software. Int. Study IMP PŁ, Łódź, in Polish (2011)
25. Żywica, G., Bogulicz, M., Czoska, B., Breńkacz, Ł.: Testing of new numerical model of gas bearings for different values of rotational speed and load. Int. Study IMP PAN, No. Arch. 749/2011, Gdańsk, in Polish (2011)
26. Howard, S.A., San Andres, L.: A new analysis tool assessment for rotordynamic modelling of gas foil bearings. ASME Turbo Expo 2010: Power for Land, Sea and Air GT2010, Glasgow (UK), 14–18 June 2010 (GT2010-22508)
27. Della Corte, C., Radil, K.C., Brucker, R.J.: Howard S.A.: Design, fabrication and performance open source generation I and II compliant hydrodynamic gas foil bearings. NASA/TM-2007-214691, April 2007
28. Żywica, G., Bogulicz, M., Czoska, B., Kreft, J.: Verification of the foil bearing model using the results of experimental studies. Int. Study IMP PAN, No. Arch. 833/2012, Gdańsk, in Polish (2012)
29. Żywica, G., Bogulicz, M., Kiciński, J., Czoska, B., Bagiński, P., Breńkacz, Ł.: Analysis of the dynamic properties of rotor-bearing system for different constructional variants of foil bearings, using developed numerical models. Int. Study IMP PAN, No. Arch. 137/2013, Gdańsk, in Polish (2013)
30. Kiciński, J., Żywica, G., Bagiński, P., Bogulicz, M., Czoska, B.: Analysis of the dynamic properties of the rotors supported on gas foil bearings, designed to use them on the test rig. Int. Study IMP PAN, No. Arch. 986/2013, Gdańsk, in Polish (2013)
31. Kiciński, J., Żywica, G., Bogulicz, M., Czoska, B.: Simulation studies to assess the impact of initial clamp on the dynamic properties of the rotor with foil bearings. Int. Study IMP PAN No. Arch. 392/2013, Gdańsk, in Polish (2013)
32. New Hampshire Ball Bearings, Inc.: Miniature and Instrument Ball Bearing. Product and Engineering, Catalogue of the Company NHBB (2012)
33. NSK Motion & Control, Inc.: Super Precision Bearings. Catalogue of the Company NSK
34. New Hampshire Ball Bearings, Inc.: Roller and Ball Bearings. Catalogue of the Company NHBB
35. New Hampshire Ball Bearings, Inc.: Hi Tech Engineering. Catalogue of the Company NHBB
36. Mohawk Innovative Technology, Inc.: Zero Clearance Auxiliary Bearing Developed, vol. 5. January 2005
37. Bagiński, P., Żywica, G., Kiciński, J.: Experimental studies of the dynamic properties of the microrotor with ceramic rolling bearings. Int. Study IMP PAN, No. Arch. 431/2013, Gdańsk, in Polish (2013)
38. http://www.utrzymanieruchu.pl
39. http://www.skf.com
40. http://www.designnews.pl
41. Kingsbury, Inc.: Kingsbury Magnetic Bearings. Brochure available at the website of Kingsbury Company
42. http://www.kingsbury.com
43. http://www.westwind-airbearings.com

Chapter 4
Modelling and Analysis of Vibrations of Microturbines

Abstract This chapter is devoted to the issues of microturbine modelling and analysis of their dynamic properties. The proposed methods of analysis of microturbines' individual components are discussed on example of two microturbine prototypes, which were designed in the IMP PAN. Radial and axial microturbine were analyzed, both with capacity approx. 2.5 kW. For those microturbines the following analyses were carried out: vibration of the rotor-bearing system for various bearing systems, modal analysis of the rotor and the frame, strength analysis, kinetostatic stress analysis and thermal stress analysis. The impact of selected design and operating parameters on the dynamic properties of microturbines was examined by computer simulation methods. During the simulation studies discussed in this chapter both our own and commercial software were used, and the verification of numerical models was based on the results of experimental studies.

This chapter will discuss simulation tests of two variants of energy microturbines developed in parallel: radial and axial microturbine. The numerical models applied during the analysis of bearings and rotors successfully passed experimental verification, and the results obtained were used to refine the design of microturbines, and were helpful in interpreting the results of experimental research. Due to the large number of analyses conducted and the volume of results obtained, this chapter presents only selected, the most interesting effects of the studies carried out. Interested readers will be able to find more details in the cited works.

As part of the ongoing research project, microturbines with electricity power of approx. 2.5 kW were developed, created on the basis of the two alternative concepts developed by two independent research teams. The team from the Technical University of Łódź under the supervision of Prof. Z. Kozanecki developed a variant of a high-speed radial microturbine. At the same time, the team of Prof. K. Kosowski from Gdańsk University of Technology worked on developing a low-speed axial microturbine.

For each of these variants of microturbines, the analysis of the rotor-bearing system was analyzed for several alternative bearing systems, including: slide bearings greased with a low-boiling medium, gas bearings, foil bearings and rolling

J. Kiciński and G. Żywica, *Steam Microturbines in Distributed Cogeneration*,
DOI 10.1007/978-3-319-12018-8_4

bearings. Additionally, the following analyses were conducted: modal, strength, thermal loads, and the impact of the stiffness of supports on the dynamic characteristics of the rotor. Calculations were based on the geometry of microturbines and design parameters of analyzed fluid-flow machines developed by constructors.

4.1 Radial Microturbine

The full characteristics of the developed structural solution of the radial microturbine have been provided in the works of the team of Kozanecki [1, 2]. This was a radial turbine, four-stage (two centripetal stages and two centrifugal stages), for which the nominal speed was 23,800 rpm. The external diameter of the rotor blade amounted to approx. 70 mm at a rotor length close to 260 mm. Mounted between the bearings was a sleeve of an electricity generator. The bearing system in the first version of the microturbine consisted of two transverse and two longitudinal gas bearings (aerostatic), powered by steam of a low-boiling medium. The first version of the microturbine was developed exclusively for laboratory tests and, therefore, was equipped with a set of additional measurement sensors. Preliminary studies of the first variant of the construction demonstrated the need to implement a few structural amendments. The most important change concerned the shape of channels leading the working medium to the turbine blades. The diameter of the channel was increased and it was better profiled [2]. In the new solution, fresh steam was fed to the turbine through two stub tubes. The second, improved version of the microturbine (Fig. 4.1), integrated with a mobile test rig, was prepared for testing on the target working medium of a micro power plant in the laboratory of IMP PAN.

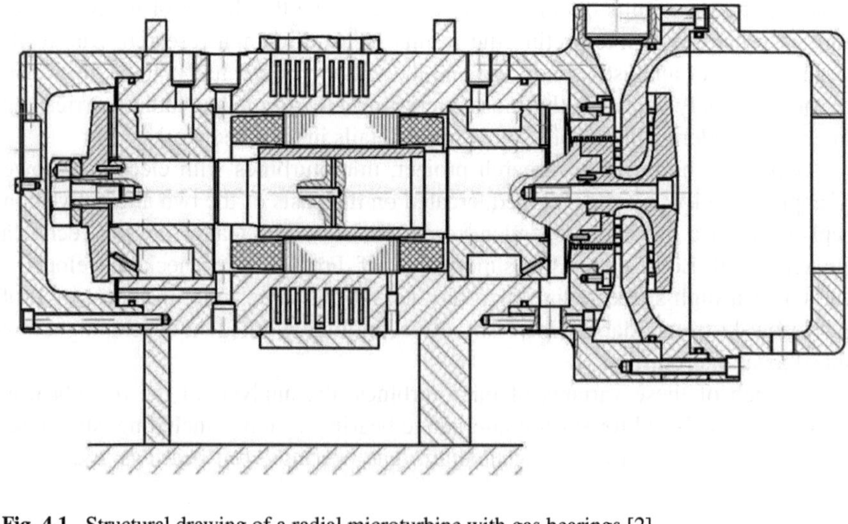

Fig. 4.1 Structural drawing of a radial microturbine with gas bearings [2]

In the third variant of the turbine generator, aside from minor structural changes, a new innovative bearing system was using aerodynamic foil bearings [3]. Bearings of this type are able to work properly without additional supply of the lubricant medium. This greatly simplifies the construction of the microturbine, because it does not require the use of a system of supplying the bearings. Additionally, the use of these bearings enables faster start-up of the rotor (without the procedure of heating the bearings), and work at higher rotational speeds. However, such bearings have certain disadvantages, which are discussed in Sect. 5.2. In this variant of microturbines, the high-speed generator was integrated with the shaft. A generator was used that was synchronous with fixed magnets, for which the maximum power amounts to 8.4 kW at a speed of 80,000 rpm. The next sections of the work show the most important results of simulation calculations.

4.1.1 Analysis of Rotor with Slide Bearings

This section discusses the analysis of the rotor of a radial microturbine with hydrodynamic bearings, in which the lubricant medium was a low-boiling medium in liquid form. Studies were conducted using the programs of the system MESWIR. Figure 4.2 demonstrates the FEM model of a four-stage radial microturbine. Calculations for the rotor fitted with slide bearings were carried out in the scope of rotational speeds from 2,000 to 30,000 rpm [4]. The size of unbalance was chosen based on the standard ISO 1940. After many attempts, bearings of the following dimensions were proposed as optimal in terms of the system's dynamics:

- Bearing 1: diameter of journal—24 mm, width of sleeve—12 mm, radial clearance—18 μm;
- Bearing 2: diameter of journal—34 mm, width of sleeve—12 mm, radial clearance—25 μm.

The diameters of bearing journals were not subject to modifications as their dimensions were chosen by the developers of microturbines. In order to improve

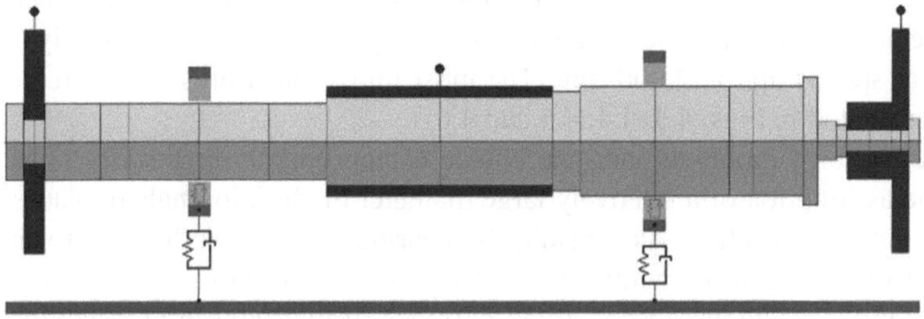

Fig. 4.2 The FEM model of the radial microturbine rotor with slide bearings developed in the system MESWIR

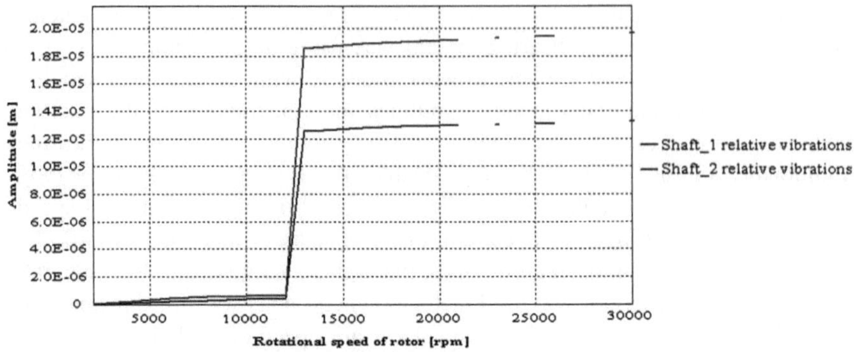

Fig. 4.3 The course of amplitudes of journal-sleeve relative vibrations for the radial microturbine in the scope of rotational speeds from 2,000 to 30,000 rpm

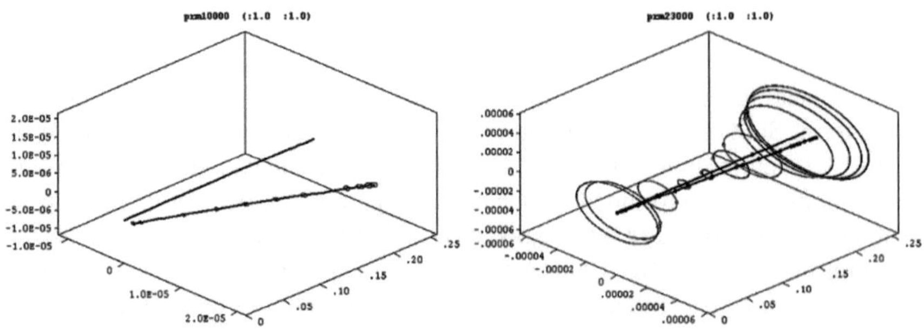

Fig. 4.4 The shape of the rotating shaft line of a radial microturbine at a speed of 10,000 and 23,000 rpm

the dynamic properties of the rotor with slide bearings, in relation to the project (with gas bearings) the width of bearings and size of the oil clearance were changed. Despite strenuous efforts which aimed to propose a bearing system that allowed stable performance throughout the analyzed scope of rotational speeds, slide bearings lubricated with a liquid low-boiling medium that would meet these requirements could not be selected. The analyzed rotor worked stably only in terms of speeds up to 12,000 rpm. The most important results of the studies have been presented in Figs. 4.3, 4.4, 4.5 and 4.6.

The cause of problems with the choice of appropriate dimensions of bearings was the assumption of a relatively large diameter of shaft journals in places where the bearings were fitted. As a result, the bearings were underloaded. Even a significant reduction in the width of the sleeve did not improve the situation and in both bearings, at higher speeds, the occurrence of hydrodynamic instability could be observed (dominant component x/2 in Fig. 4.6). However, it should be noted that the proposed construction of the radial microturbine rotor has been developed

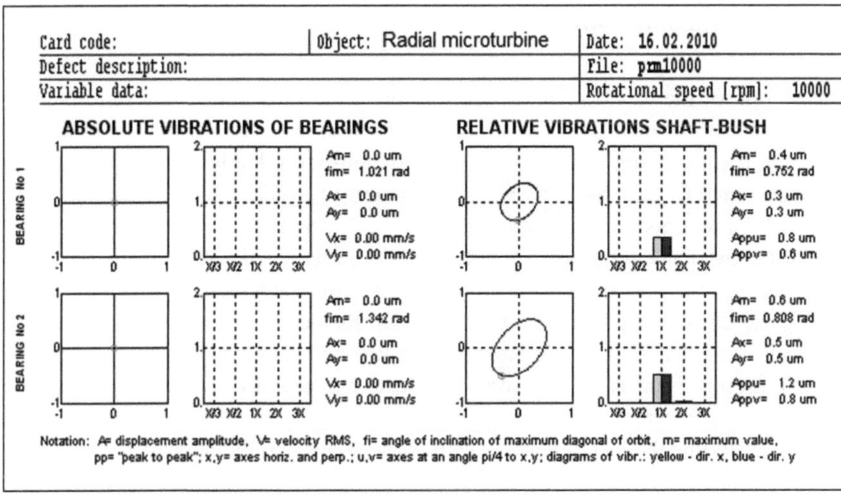

Fig. 4.5 Diagnostic card for a radial microturbine at a speed equal to 10,000 rpm

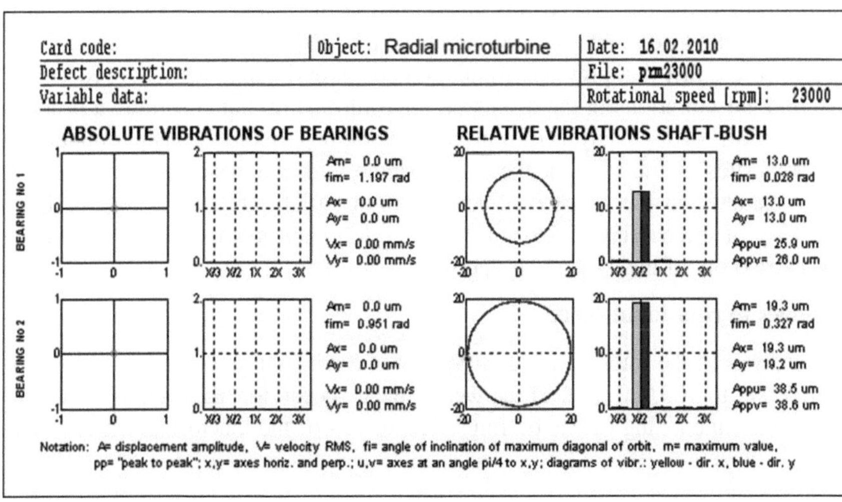

Fig. 4.6 Diagnostic card for a radial microturbine at a speed equal to 23,000 rpm

for the purpose of carrying out tests at a laboratory workstation, and there is a possibility of reducing the diameter of the rotor in the target structural solution in order to improve dynamic stability. Additionally, in accordance with the results of experimental and simulation tests (presented in the next section) the rotor of the tested microturbine with gas bearings performed steadily throughout the entire scope of rotational speeds, therefore, eventually there was no need to use hydrodynamic bearings.

4.1.2 Analysis of Rotor with Gas Bearings

The second of the tested systems was a rotor of the same radial microturbine with self-operating gas bearings. Therefore, the analysis was carried out for practically the same rotor-bearing system as in the actual structural solution of the microturbine [2]. Therefore, these tests were conducted in a wider scope than e.g. in the case of the previously discussed hydrodynamic bearings.

During the tests programs included in the system MESWIR were used. The rotor model was built from 23 beam elements with 4 degrees of freedom in each node. Discretisation of the model performed for the needs of the kinetostatic analysis has been presented in Fig. 4.7. The tested rotor had two journals with a diameter of 34 mm, prepared for using transverse bearings, and two discs enabling the assembly of longitudinal bearings. The developed rotor model reflected the real geometry of the system, excluding the insignificant construction details and miniature rotor blades.

The lubricant medium of bearings during the performed calculations was air. The diameter of the bearing sleeves resulted from the diameter of journals and amounted to approx. 34 mm, at a width of 40 mm. The base oil clearance for this bearing was determined at the level of 0.025 mm. Based on the operational parameters of microturbines, the base speed was determined at the level of 25,000 rpm (approx. 417 Hz). Static load of bearings resulted from the load of the shaft and discs.

Due to the fact that previous simulation and experimental tests showed that the longitudinal vibrations in the discussed microturbine do not constitute a problem, the focus was on the analysis of transverse vibrations. In order to better understand the properties of the rotor of the microturbine with gas bearings, the tests were first conducted in the scope of kinetostatics at variable rotational speeds, loads, geometry of the bearing and viscosity of the lubricant medium. Then an analysis was carried out of the dynamic properties.

Changes in rotational speed

The analysis of the rotor of a microturbine was carried out in the scope of the speed from 7,000 to 100,000 rpm. These were typical speeds for microturbines, enabling the proper functioning of gas bearings. The results of the kinetostatic analysis have

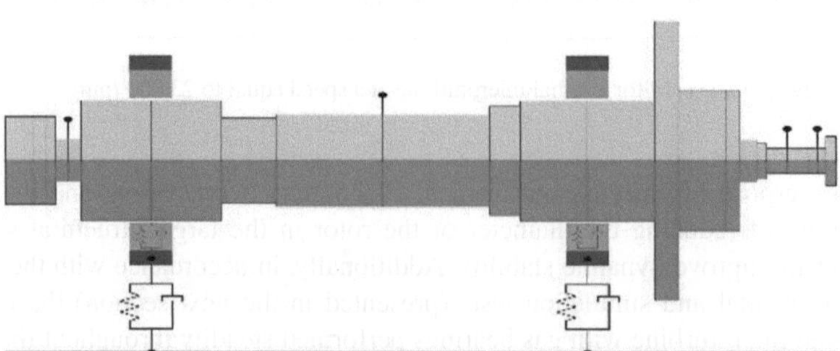

Fig. 4.7 The FEM model of the rotor of a radial microturbine with gas bearings

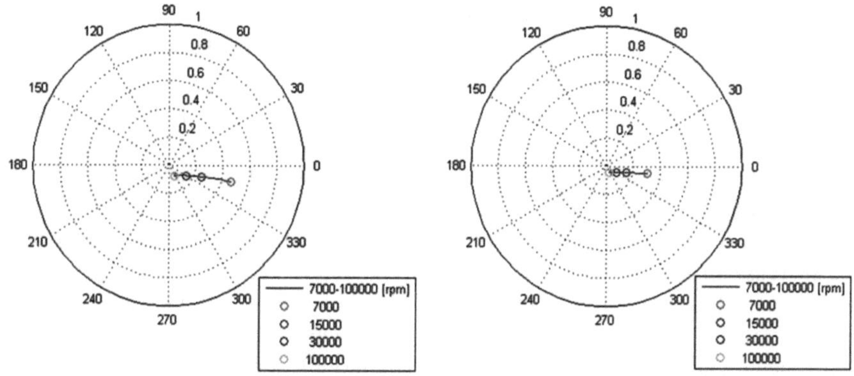

Fig. 4.8 Change in the position of the centre of journals in the microturbine's rotor bearings caused by a change in speed

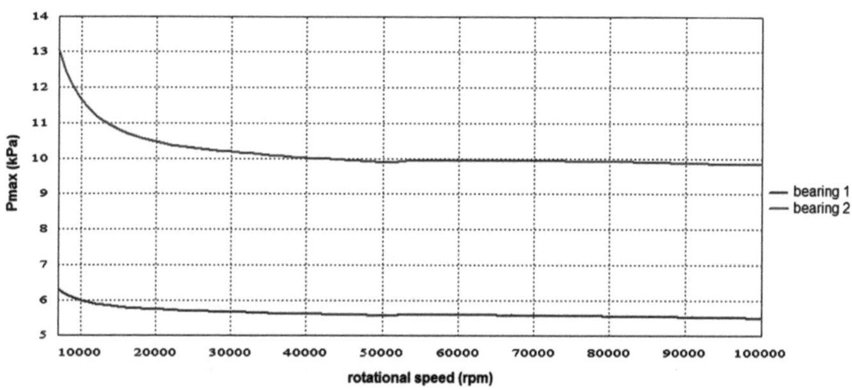

Fig. 4.9 Change in the maximum value of pressure in the clearance gap of bearings of the microturbine depending on speed

been presented in the form of: changes in the position of the journal centre, the course of maximum pressure in the bearing, the distribution of pressures and coefficients of stiffness and damping of bearings. Because the rotor of the microturbine was asymmetrical, for each of the bearings different results were obtained. That is why the characteristics of the bearings have been presented separately for each bearing.

Figure 4.8 presents changes in the position of bearing journals depending on the rotational speed. Along with the increase in speed, the centre of the journal moved toward the centre of the sleeve. The bearing situated near the rotor disc worked with a greater eccentricity, which was visible particularly at lower speeds. The maximum pressure in bearings depending on the speed has been shown in Fig. 4.9. Together with the increase in speed, it decreased; however, a drop in pressure over 30,000 rpm was almost invisible. Because bearing no. 2 situated by the rotor disc was more loaded, high pressure occurred in it.

Despite the significant differences in maximum pressure, the pressure distribution in the two bearings was very similar (Fig. 4.10). In the lower part of the sleeve there occurred an area of increased pressure (lubricating wedge), while in the upper part, an area of reduced pressure. Mild changes in pressure occurred between these areas.

The next determined characteristics were the courses of coefficients of stiffness and damping of bearings. They have been presented in Figs. 4.11, 4.12, 4.13 and 4.14. From the presented courses of coefficients of stiffness it appears that with the increase in speed, the changes of this parameter were very small. The stiffness of bearings gradually decreased to a speed of approx. 40,000 rpm, and then it was almost constant. The main stiffness coefficients achieved values that were a few times greater than diagonal stiffness. The more loaded bearing no. 2 at lower speeds was characterized by greater stiffness, but in the remaining scope of the speed had very similar stiffness characteristics.

Similar trends were observed in the case of damping coefficients, which throughout the entire tested scope had a relatively low value. After exceeding 70,000 rpm, these coefficients almost did not change. Very low damping coefficients are typical of aerodynamic slide bearings.

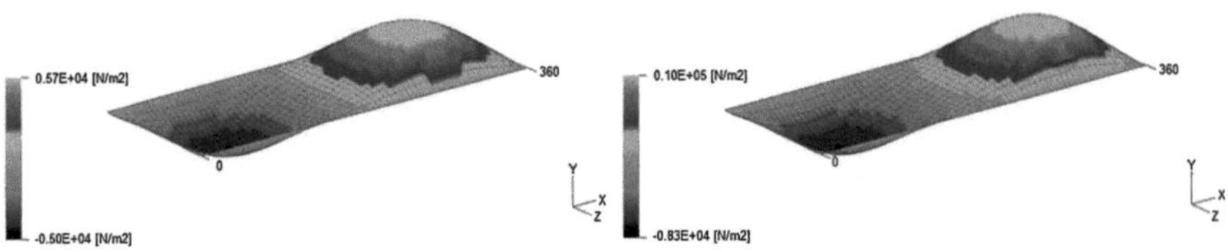

Fig. 4.10 Pressure distribution in gas bearings of the microturbine at a speed of 25,000 rpm

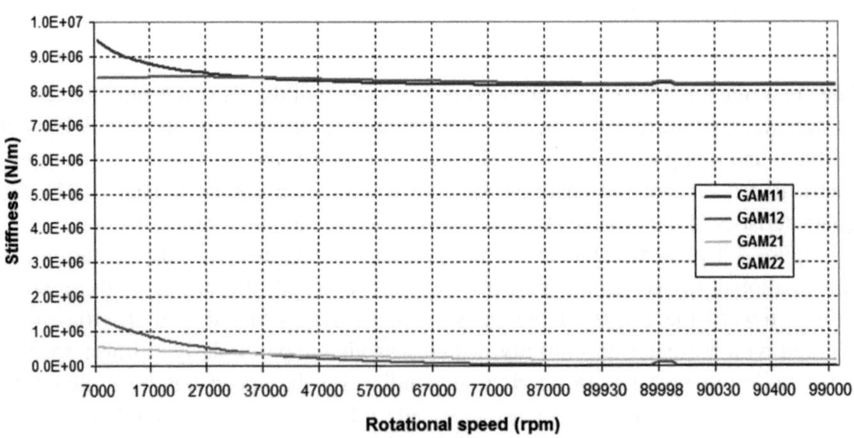

Fig. 4.11 Change of stiffness coefficients in gas bearing no. 1

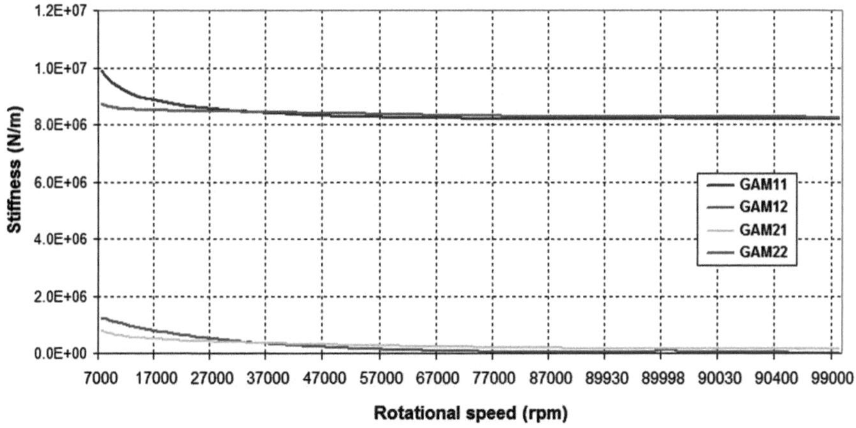

Fig. 4.12 Change of stiffness coefficients in gas bearing no. 2

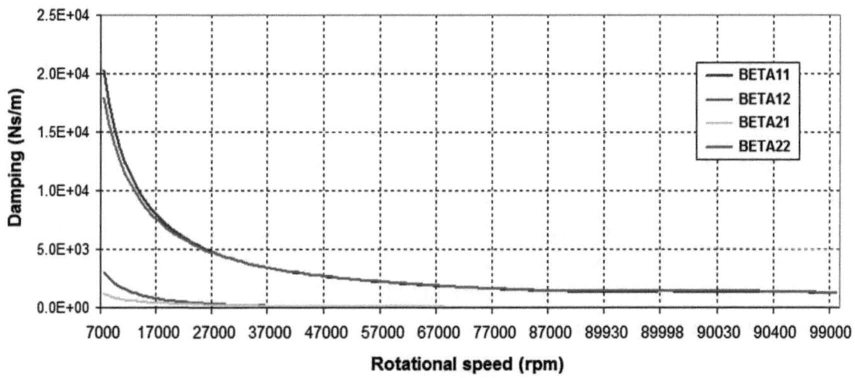

Fig. 4.13 Change of damping coefficients in gas bearing no. 1

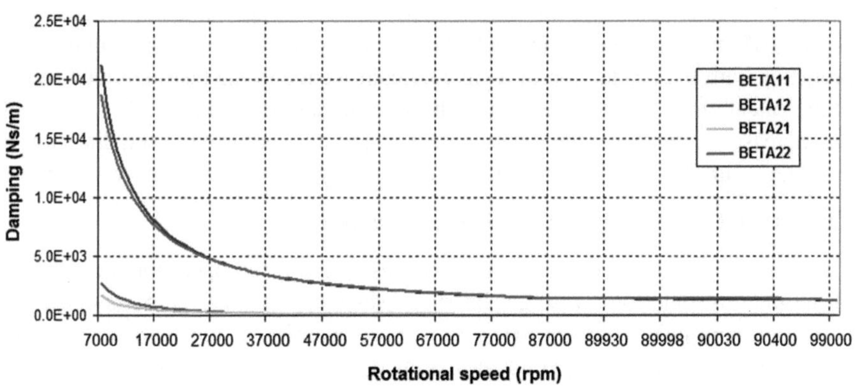

Fig. 4.14 Change of damping coefficients in gas bearing no. 2

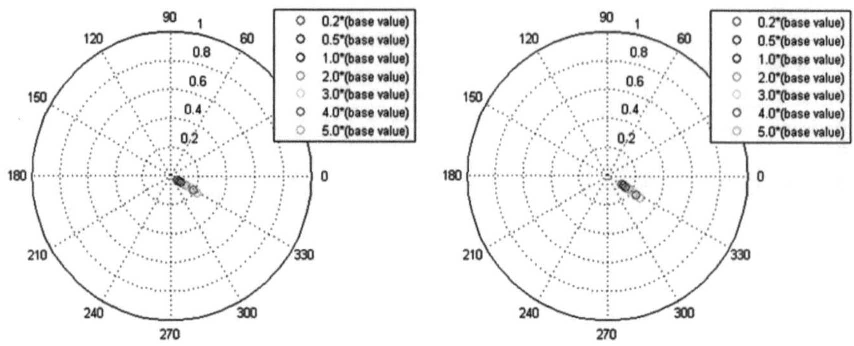

Fig. 4.15 Change in the position of the centre of the journals in the bearings of the microturbine's rotor caused by a change in load

Load changes

The analysis of load changes in the case of the microturbine rotor was carried out by changing the load of all discs. The base values were as follows: disc 1–2.954 N, disc 2–1.442 N, disc 3–0.415 N, disc 4–1.469 N. A few cases were tested, for which the loads of discs changed in the range from 20 to 500 % of the base value. The results of calculations have been presented graphically, showing the change in the position of the centre of journals (Fig. 4.15). Increasing the load of the rotor, thus the bearings themselves, caused the displacement of journals toward the surfaces of the sleeve along the semi-circle of static equilibrium. Interestingly, the quite large, fivefold increase in the load caused a relatively small change in the position of the journals. It can be concluded that this indicated a large reserve of the bearings' carrying capacity.

Changes in the geometry of the bearing

Modification of the geometry of the tested gas bearings took place by changing the size of the oil clearance. The base value was 0.025 mm, and the tests were conducted for values ranging from 0.01 to 0.10 mm. At higher values of oil clearance (from 0.05 mm), no converging calculations were obtained, which resulted from problems with the production of a gas lubricant wedge in the bearings. The results of the calculations for the two bearings have been included in Fig. 4.16. Along with the increase in clearance, the displacement of journals occurred in the direction of the sleeves, until eccentricity at the level of 0.4 was achieved in one of the bearings.

Changes in the viscosity of the lubricant medium

The analysis of the effect of the medium viscosity on the operational conditions of the microturbine rotor with gas bearings was conducted on the basis of air viscosity amounting to 0.185×10^{-4} Pa s. The viscosity of the lubricant medium during the analysis of this parameter assumed values ranging from 10 to 1,000 % in relation to the base value. Cases for which it was not possible to obtain convergent solutions were not included in the results shown in Fig. 4.17. Such a situation

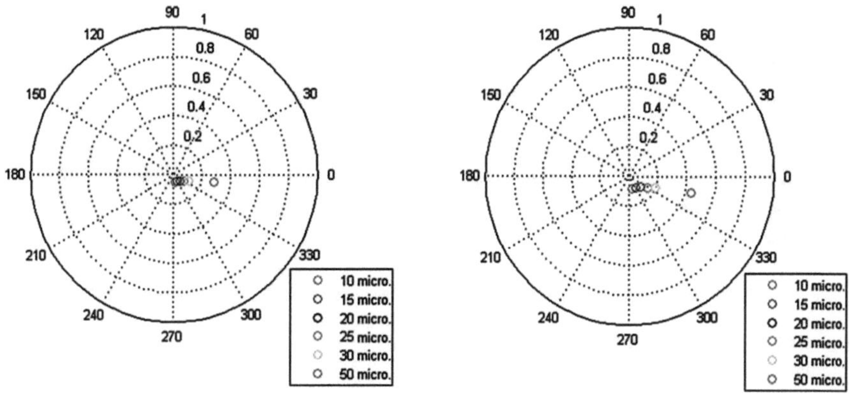

Fig. 4.16 Changes in the location of the centre of journals of gas bearings depending on radial clearance

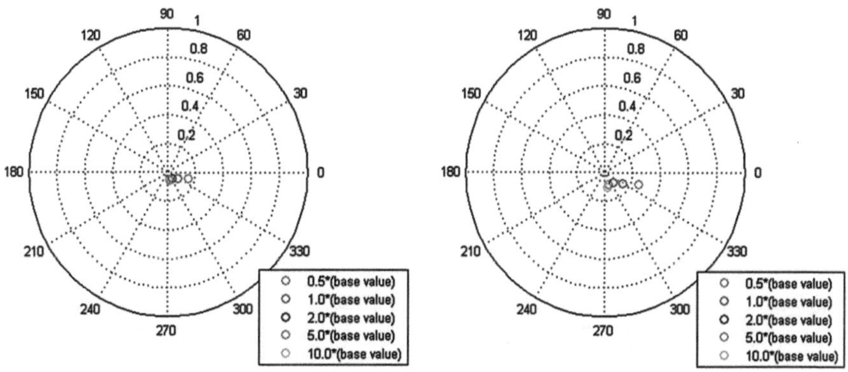

Fig. 4.17 Changes in the location of the centre of journals of gas bearings depending on the viscosity of the lubricant medium

occurred at a viscosity lower than half the base value. The obtained results showed that the increase in viscosity caused the displacement of the journal toward the centre of the sleeve, just as it did in the case of the theoretical rotor. This was associated with better conditions of lubricating bearings and the formation of a lubricant wedge of greater thickness.

The results of the kinetostatic calculations discussed above for various operating conditions of the rotor and bearings of the microturbine proved proper functioning of the numerical model used, even in the case of quite a complex construction.

Analysis of the dynamic properties

The next stage of the test calculations of the radial microturbine's rotor with gas bearings, was the analysis of the dynamic properties (forced vibrations of the system). In this case, it was also assumed that the bearing system consisted of two aerodynamic

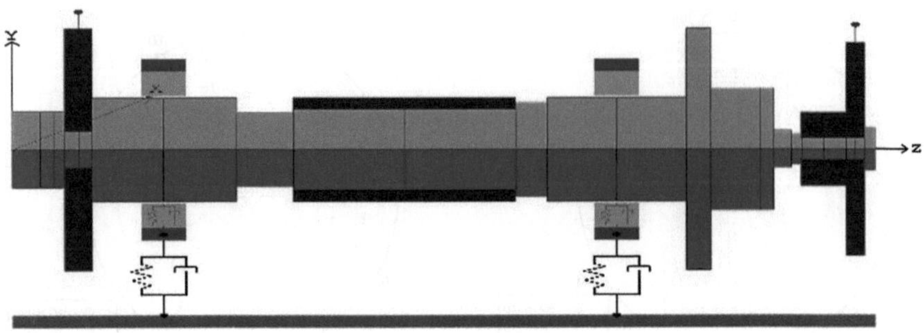

Fig. 4.18 The FEM model of the radial microturbine's rotor with gas bearings used to carry out calculations in the scope of dynamics

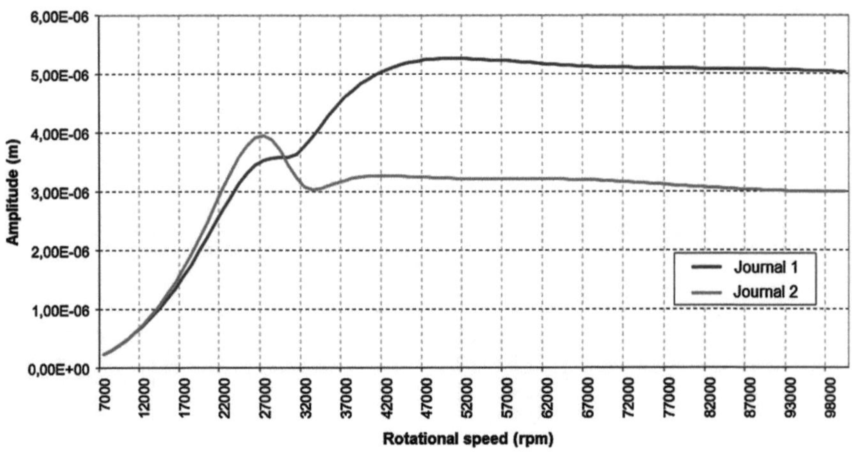

Fig. 4.19 Amplitudes of vibrations of journals of the microturbine rotor with gas bearings

gas bearings. The discretization of the model used in the dynamic calculations has been presented in Fig. 4.18. The journal and disc no. 1 were located on the left side of the shaft, as shown in Fig. 4.18. In these studies the focus was on the analysis of transverse vibrations.

Figures 4.19 and 4.20 present the vibrations of two bearing journals and three discs visible in Fig. 4.18 in the speed range from 7,000 to 100,000 rpm. Unbalance was applied to the two extreme discs and amounted to 4.5 g mm for each of them. In addition, angular displacement was applied of two unbalanced loads by 180°, which aimed to induce so-called conical forms of vibrations.

The analysis conducted shows that the greatest increases in vibration amplitude were observed at a speed of approx. 27,000 rpm and approx. 50,000 rpm. At 27,000 rpm for the tested system, so-called conical vibrations occurred, which was mainly displayed through increased vibrations of the more heavily loaded end of

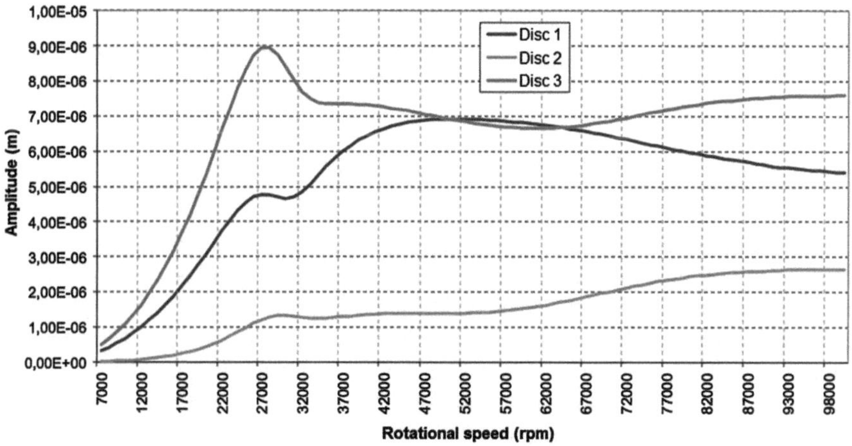

Fig. 4.20 Amplitudes of vibrations of discs of the microturbine rotor with gas bearings

the shaft (journal 2 and disc 3), but they were also visible for other elements. At approx. 50,000 rpm a similar shape of the rotating shaft could be observed (conical form), but also a small deflection of the shaft occurred in the necking between bearing 2 and disc 3.

Since the amplitudes of vibrations were very small (they reached only a few μm), even a slight deflection of the shaft resulted in their growth, which can be seen especially at the two ends (disc 1 and 3 and journal 1). At higher speeds stable operation of the system could be observed with small changes in the level of vibrations. The amplitudes of vibrations in bearings at nominal speed (approx. 25,000 rpm) were less than 4 μm (in bearing no. 2 they were approx. 10 % greater). At higher speeds, the amplitude of vibrations in bearing no. 1 exceeded 5 μm. For discs in the entire analyzed range, the largest amplitude of vibrations reached up to 9 μm (disc 3).

The results of calculations obtained can be more or less compared with the results of experimental studies discussed in the work [5]. Experimental studies were carried out at a load of the generator and air flow through the turbine, which was not included in the simulation calculations. The main difference, however, concerned gas bearings, as in an actual machine these were powered gas bearings (aerostatic). Respectively, the results of the measurements can be treated as approximate, constituting only a point of reference. Experimental studies have shown that during the acceleration of the machine up to 25,000 rpm, the amplitudes of vibrations changed to a minor extent, which could also have been the result of the small accuracy of the measuring apparatus. The performance of tests on measurement paths of a large diameter (on the edge of the discs), as well as disturbances of the magnetic field through the functioning of a generator with permanent magnets were provided as the cause of the low accuracy of measurements. The peak-to-peak amplitude of vibrations measured at bearing 1 reached the value of approx. 2 μm, and at bearing 2 approx. 9 μm. These were similar values to those obtained through computer simulation using the developed numerical model.

4.1.3 Analysis of Rotor with Foil Bearings

The simulation tests of the radial microturbine with foil bearings, discussed in this section of the book, concerning lubricant bearings with a low-boiling medium in liquid form. These tests were conducted using programs from the package MESWIR and ABAQUS. During the calculations, a newly developed, proprietary model of foil bearings was used, taking into account the interactions that occur between the fluid and solid body. Similar analyses, also concerning foil bearings but lubricated by gas mediums, have been presented earlier in Sect. 3.2.

Changing gas bearings for foil bearings required certain modifications in the construction of microturbines [3]. These changes have been presented in Fig. 4.21.

The rotor model was developed according to the standards adopted in the programs from the MESWIR series and consisted of 23 beam elements with 6 degrees of freedom at each node. The discretization of the model has been presented in Fig. 4.22. The developed rotor model reflected the geometry of a real system without rotor blades, which dynamic analysis is discussed later in this chapter.

During the analysis the residual unbalance of the rotor was taken into account, which did not exceed acceptable values for this class of machines. On the two extreme discs, unbalance with a value of 0.35 g mm was assumed, and on the middle disc (the generator sleeve) imbalance with a value of 0.34 g mm rotated by 180°.

The model of the entire foil bearing consisted of a fluid-flow and structural load bearing layer. The analysis of the fluid-flow lubricant medium in the oil clearance was performed in the programs from the MESWIR series, which took place simultaneously with the analysis of the rotor. The model of the structural load bearing layer of the foil bearing was developed using the ABAQUS program, which

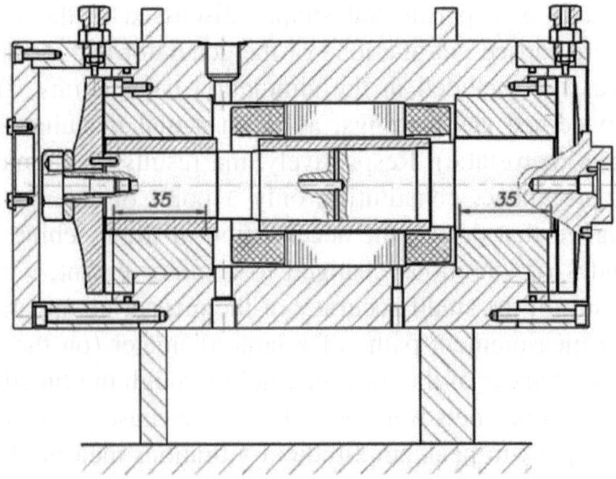

Fig. 4.21 The structural solution of the rotating system of the prototype radial microturbine with foil bearings [3]

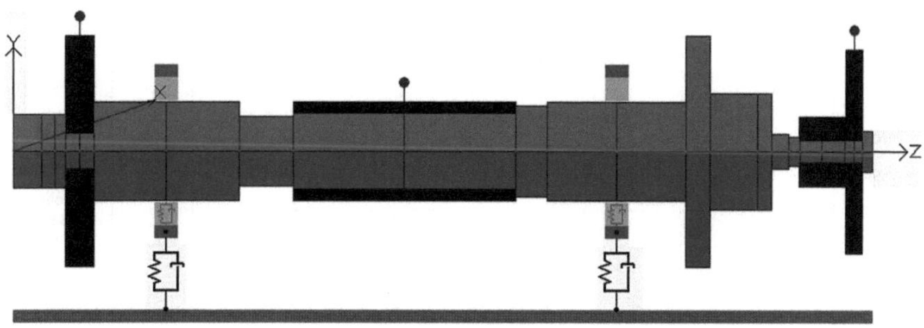

Fig. 4.22 The FEM model of the radial microturbine rotor with foil bearings was developed using programs from the MESWIR series

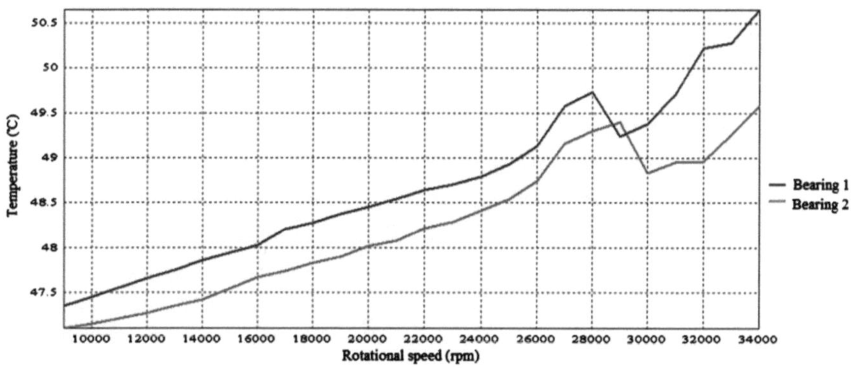

Fig. 4.23 The change in temperature of the lubricant factor of foil bearings depending on the rotational speed of the rotor

enables to determine the deformation of the system of highly-sensitive foil, taking into account the non-linear geometry and contact phenomena. Due to the small calculation performance of 3D models, a 2D model of the structural load bearing layer of the foil bearing was used in the conducted tests. The dimensions of the tested bearing were adjusted to the rotor discussed earlier. The nominal diameter of the bearing, equal to the inner diameter of the top foil, was 34 mm. The inner diameter of the bearing sleeve was 35 mm. Both foils had a thickness of 0.1 mm. The bump foil had 44 bumps with a radius of 2 mm. The thermodynamic medium of the tested microturbine, as well as the lubricant medium of bearings was a chemical compound from the group of low-boiling mediums with the marking HFE7100.

Figure 4.23 shows computationally determined temperature changes of the lubricant medium in the function of rotational speed of the rotor. The presented characteristics show that as a result of the increase in speed from 10,000 to 34,000 rpm, the temperature of the low-boiling factor increased by approx. 3 to 50.5 °C. The initial temperature (at the inlet to the bearings) was 40 °C, therefore, the total temperature rise amounted to approx. 10 °C.

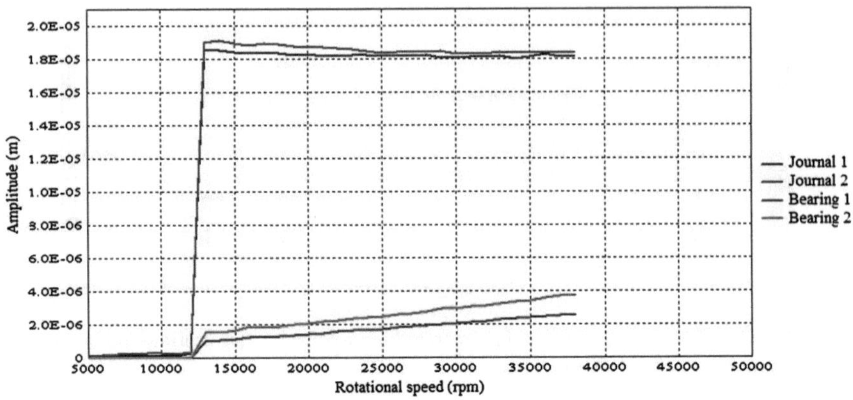

Fig. 4.24 The courses of amplitudes of relative vibrations of bearing journals and absolute vibrations of sleeves for the microturbine rotor with foil bearings

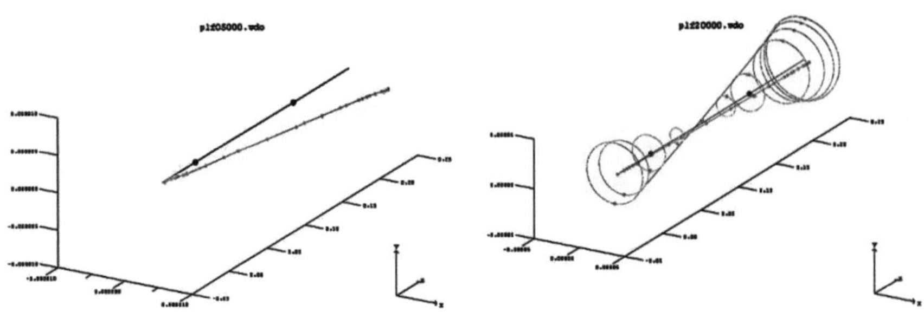

Fig. 4.25 The shape of the rotating shaft of a microturbine at a speed of 5,000 and 20,000 rpm

The following diagrams show the most important results of calculations obtained during the dynamic analysis of the microturbine rotor with foil bearings. Figure 4.24 shows the course of amplitudes of relative vibrations of bearing journals and absolute vibrations of sleeves. For the analyzed system it was not possible to obtain concurrent solutions for speeds higher than 38,000 rpm. The obtained courses of amplitudes of vibrations had a distinctive, sudden rise in value at a speed of approx. 12,500 rpm. A detailed analysis of the obtained results showed that over 12,500 rpm, the analyzed system worked unstably.

Figure 4.25 demonstrates the shape of the rotating shaft at two selected rotational speeds. This diagram shows the stable work of the system at a speed of 5,000 rpm and a clear increase in the amplitudes of vibrations of the system at 20,000 rpm. The trajectories of vibrations had a similar shape to circles, and their radius reached values that were comparable to radial clearance. Such vibrations are an example of the so-called conical vibrations, which take place at a slight deflection of the shaft. A similar form of vibrations of the rotor were obtained at different speeds, above 12,500 rpm.

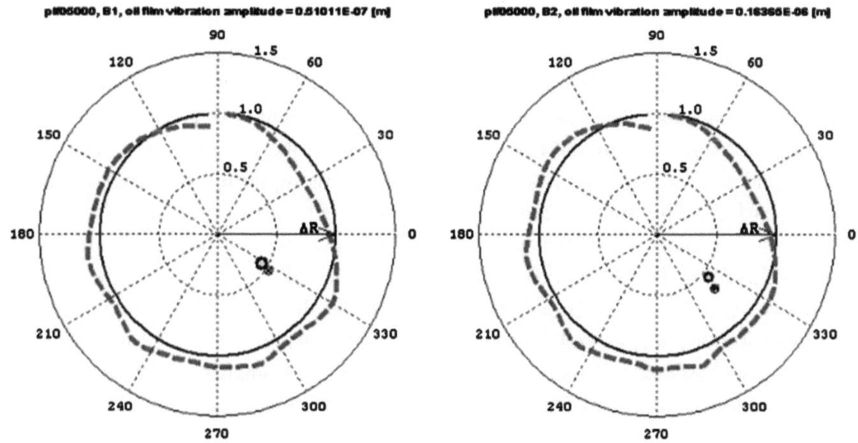

Fig. 4.26 Trajectories of vibrations of bearing journals at a speed of 5,000 rpm

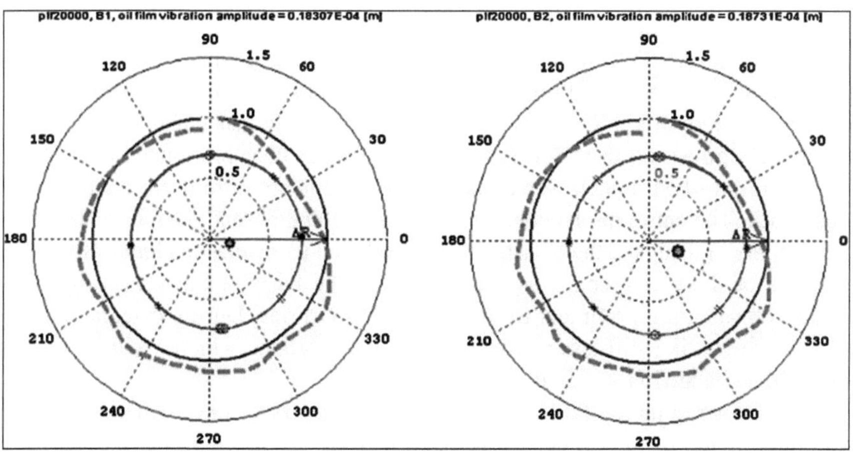

Fig. 4.27 Trajectories of vibrations of bearing journals at a speed of 20,000 rpm

The following diagrams present the trajectories of vibrations of bearing journals against the deformed clearance gap at rotational speeds amounting to 5,000 and 20,000 rpm (Figs. 4.26 and 4.27). As it has already been highlighted, at a speed of 5,000 rpm, very stable operation of the system was observed, which was shown by an almost invariable position of bearing journals. Their orbits of vibrations had a regular shape, similar to a circle with a very small radius. The situation changed greatly above 12,500 rpm, which was shown by a significant increase in vibration amplitudes of journals in both bearings. In the scope of the analyzed rotational speeds, similar deformations of the group of foils occurred. The shape of the deformed clearance gap has been marked in the diagrams below with a red dashed line.

The presentation of the obtained results in the form of diagnostic cards confirmed the fact that above 12,500 rpm the system worked unstably [6]. Therefore, the application of foil bearings in the case of the analyzed fluid-flow machine did not bring the desired effect in the form of an improvement of dynamic properties of the rotor-bearings system and the improvement of stability of operation. The tested system had similar properties as in the case of a rotor supported in classic hydrodynamic bearings. The high susceptibility of foil system and significant deformations of the oil clearance associated with this did not have a great impact on the vibrations of the rotor. The very good damping properties of the system of thin foils did not help either. As it turned out, the dimensions of the journal and sleeve were of dominant importance, as well as the underload of bearings associated with this and the very low viscosity of the lubricant factor. The combination of these two very adverse factors determined the poor dynamic properties of the system.

4.1.4 Modal Analysis of the Rotor

During the computational modal analysis, the 3D FEM model was used developed in the program ANSYS, which took into account the full geometry of the rotor and the rotating disc embedded on it [7]. This allowed to determine the form and frequency of the rotor's free vibrations and the discs and blades embedded on it. The 3D model of the rotor used during modal analysis has been presented in Fig. 4.28.

To build a FEM model, isoparametric finite elements were used, mainly eight-node ones. The entire model consisted of nearly 200 thousand degrees of freedom. During modal analysis, the bearings were modelled using elastic elements, and the stiffness of these elements was determined on the basis of the values measured in the rotor-bearing programs from the MESWIR series. During calculations, the changes in bearing stiffness were also taken into account, which occurred in relation to the change in rotational speed.

The following table (Table 4.1) presents a summary of the frequencies of free vibrations determined for not rotating rotor and rotating at a nominal speed.

Table 4.2 shows selected forms of the rotor's free vibrations, which are associated with the vibrations of the shaft and disc. Due to the very small dimensions and high stiffness resulting from this, the vibrations of individual blades occurred over 50,000 Hz.

Fig. 4.28 3D model of the radial microturbine rotor

Table 4.1 The frequencies of the radial microturbine rotor's free vibrations not rotating and rotating at a speed of 23,800 rpm

No.	0 rpm	23,800 rpm
	Frequency	
	(Hz)	(Hz)
1	1,405.1	1,422.4
2	1,405.1	1,422.4
3	1,454.2	1,507.3
4	2,081.1	2,101.3
5	2,081.2	2,101.4
6	3,420.1	3,443.0
7	4,391.6	4,400.0
8	4,392.0	4,400.4
9	5,738.0	5,744.7
10	6,332.3	6,340.3
11	6,333.1	6,341.1
12	6,971.4	6,996.7
13	6,971.7	6,996.7
14	7,282.8	6,997.1
15	8,577.9	7,291.0

Table 4.2 The base forms of vibrations of the radial microturbine rotor

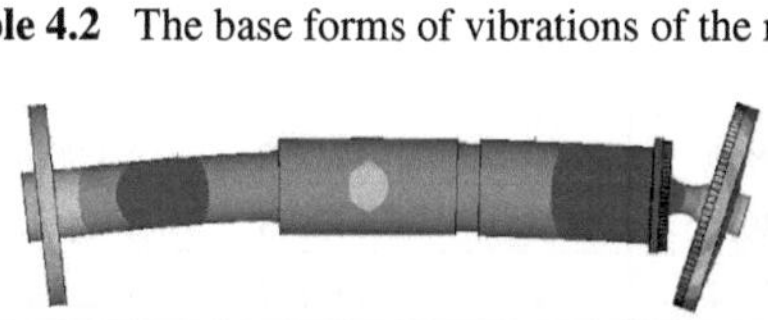	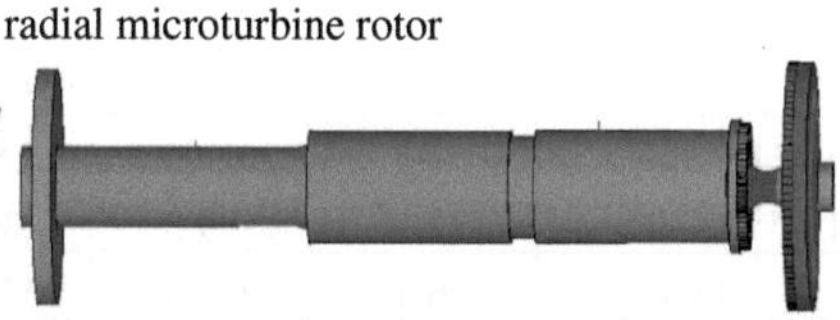
The first deflective form of vibrations of the shaft at a frequency of 1405.1 Hz	Torsional form at a frequency of 1454.2 Hz
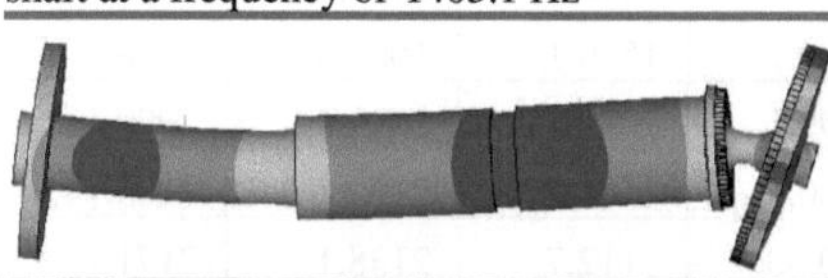	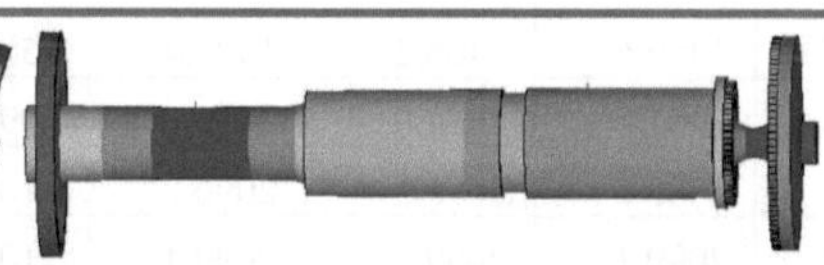
The second deflective form of the rotor at a frequency of 2081.1 Hz	Torsional form at a frequency of 3420.1 Hz
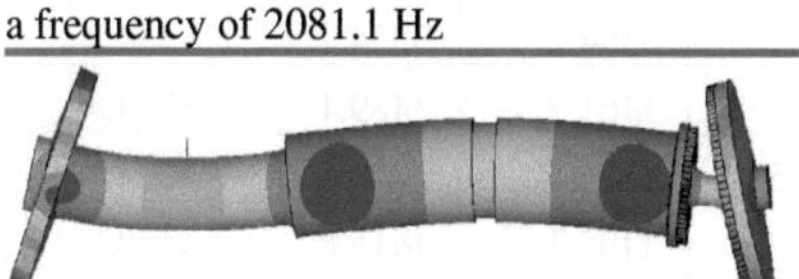	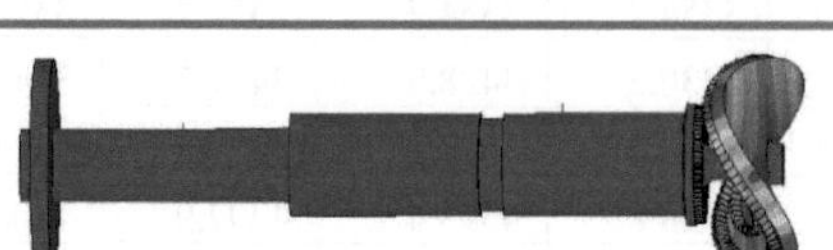
The deflective form at a frequency of 4391.6 Hz	The form of vibrations with the second node diameter on the right disc at a frequency of 6971.4 Hz

The modal analysis of the microturbine rotor showed that it is a subcritical rotor. The first form of bending vibrations of the shaft occurred at a frequency of approx. 1,400 Hz, which corresponds to a speed of 84,000 rpm, which is several times higher than the anticipated nominal speed of the microturbine. The free vibrations of the bladed disc and single blades occurred at even higher frequencies, therefore, they should not be induced during operation of the machine.

4.1.5 Modal Analysis of the Microturbine

Modal analysis of the entire radial microturbine was conducted with the full geometry of the body being taken into account [8]. During calculations, the FEM model was used consisting of over 300 thousand degrees of freedom (Fig. 4.29).

The connection of the rotor with the body, which in the actual construction is done by the bearing, has been modelled with elastic elements. The stiffnesses of bearings in the right directions have been adopted on the basis of previous calculations of the rotor-bearings system [4].

Table 4.3 shows the frequency of vibrations of a radial microturbine assembly (rotor in the body) for the full range of rotational speeds. The forms of vibrations

Table 4.3 The frequencies of free vibrations of the microturbine assembly for various rotational speeds

	rotational speed [rpm]						
No.	*2,500*	*10,000*	*20,000*	*23,800*	*30,000*	*40,000*	*50,000*
1	1139.5	1139.5	1139.5	1139.5	1139.5	1139.5	1139.5
2	1326.4	1326.4	1326.4	1326.4	1326.4	1326.4	1326.4
3	1406.2	1408.3	1417.5	1422.4	1432.2	1451.6	1474.8
4	1408.8	1409.4	1418.6	1423.5	1433.3	1452.7	1475.9
5	1459.6	1466.2	1494.4	1509.7	1540.1	1602.0	1678.2
6	2081.8	2084.0	2094.7	2099.1	2099.1	2099.1	2099.1
7	2083.2	2085.5	2096.2	2100.7	2105.7	2105.7	2105.7
8	2099.1	2099.1	2099.1	2102.0	2112.7	2138.1	2171.1
9	2105.7	2105.8	2105.9	2105.9	2114.3	2139.6	2172.6
10	3354.2	3354.2	3354.2	3354.2	3354.2	3354.2	3354.2
11	3430.5	3428.9	3441.1	3447.8	3461.2	3489.1	3524.8
12	4024.4	4024.5	4024.5	4024.5	4024.5	4024.5	4024.5
13	4394.8	4393.4	4397.9	4400.3	4405.3	4415.6	4428.8
14	4398.5	4397.2	4401.6	4404.1	4409.0	4419.3	4432.5
15	4636.5	4636.3	4636.3	4636.3	4636.3	4636.3	4636.3
16	4785.3	4784.7	4784.7	4784.7	4784.7	4784.7	4784.7
17	5731.9	5733.0	5736.5	5738.5	5742.5	5750.7	5761.2
18	5879.6	5879.5	5879.5	5879.5	5879.5	5879.5	5879.5
19	6148.5	6148.5	6148.5	6148.5	6148.5	6148.5	6148.5

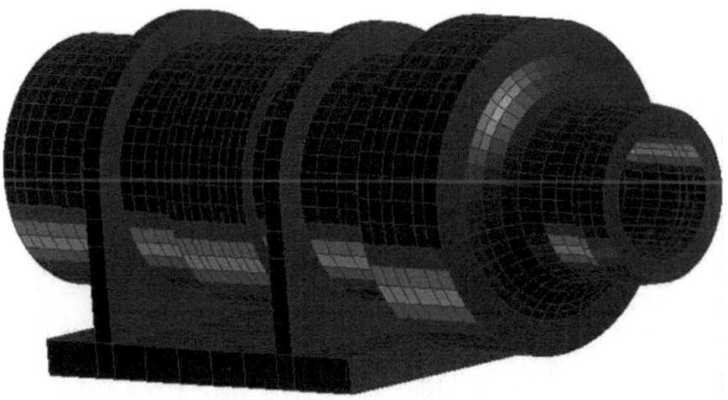

Fig. 4.29 FEM model of the radial microturbine

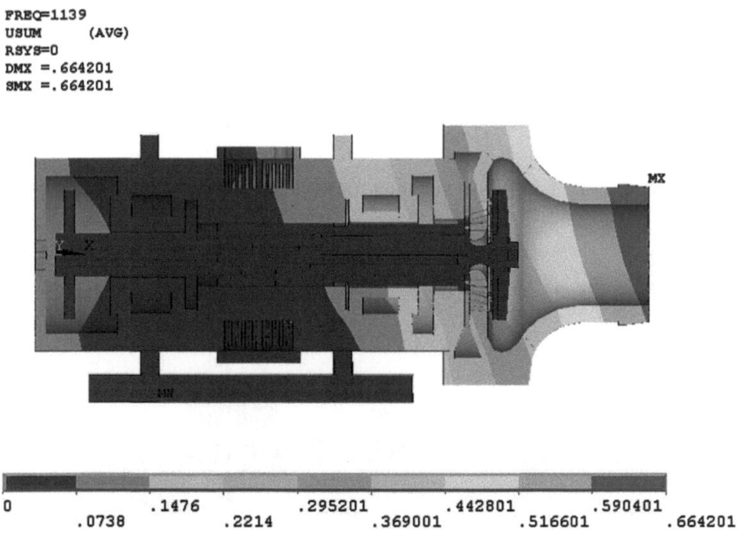

Fig. 4.30 The first form of vibrations of the radial microturbine body at a frequency of 1,139 Hz

associated with the body have been marked in black, deflective vibrations in blue, torsional vibrations in green, and longitudinal vibrations of the rotor in red.

Figures 4.30, 4.31, 4.32, 4.33, 4.34 and 4.35 show a visualization of selected forms of free vibrations of the microturbine assembly consisting of a body, rotor and bearings for the chosen rotational speed n = 10,000 rpm. With the assumed bearing model no clear couplings occurred between the rotor and the body. Along with the increase in the rotational speed, the "shaft" frequencies increased insignificantly at the same forms of vibrations. The obtained results of the simulation modal analysis demonstrated that the free vibrations in the tested machine should not be a problem. The first form of deflective vibrations of the rotor occurred at a frequency of approx. 1,400 Hz, while the nominal rotating frequency of the shaft is approx. 400 Hz. The forms of free vibrations of the body also occurred outside the scope of operation of the machine. For

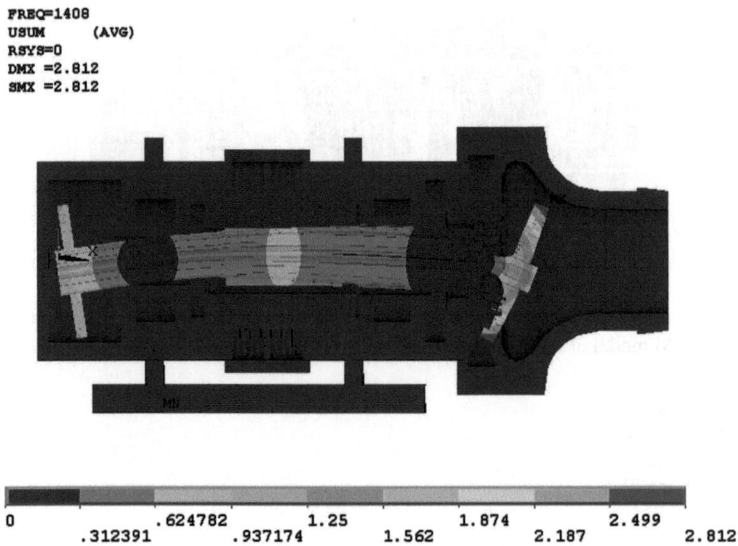

Fig. 4.31 The first form of deflective vibrations of the microturbine rotor at 1,408 Hz at a speed of 10,000 rpm

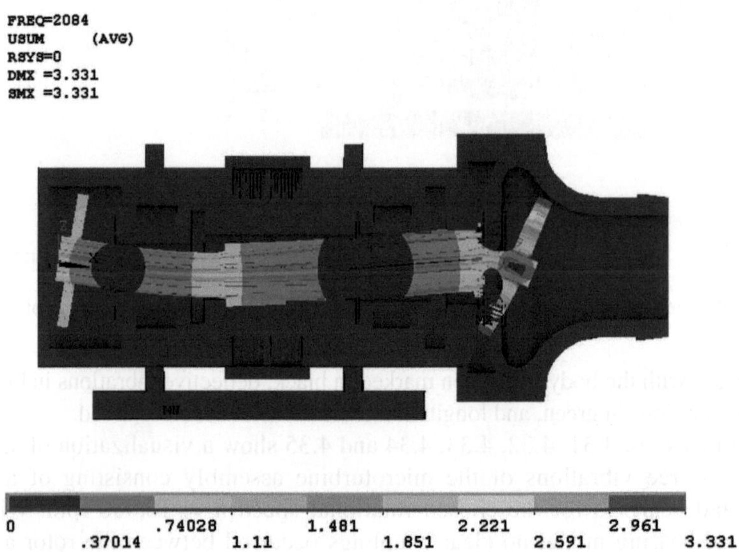

Fig. 4.32 The second form of deflective vibrations of the microturbine rotor at 2,084 Hz at a speed of 10,000 rpm

the first form of free vibrations of the body occurred at approx. 1,140 Hz and primarily concerned the oscillation of the outlet of the body, which in an actual machine will be stiffened once it is connected with the installation of the ORC micro power plant.

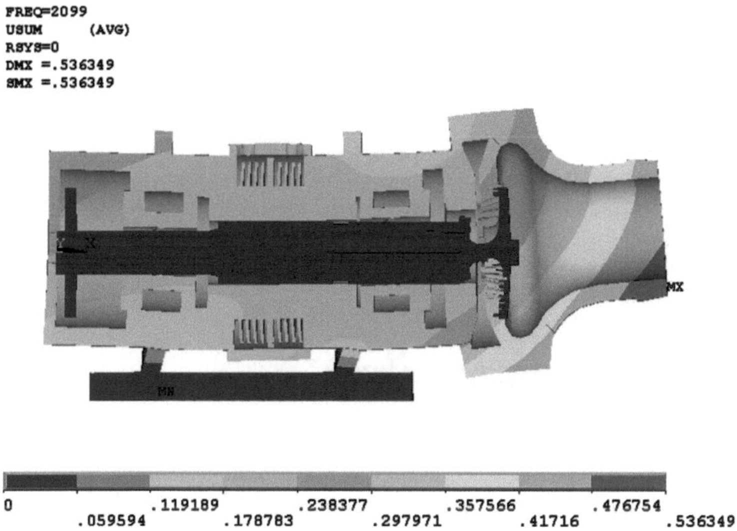

Fig. 4.33 The third form of vibrations of the radial microturbine body at a frequency of 2,099 Hz

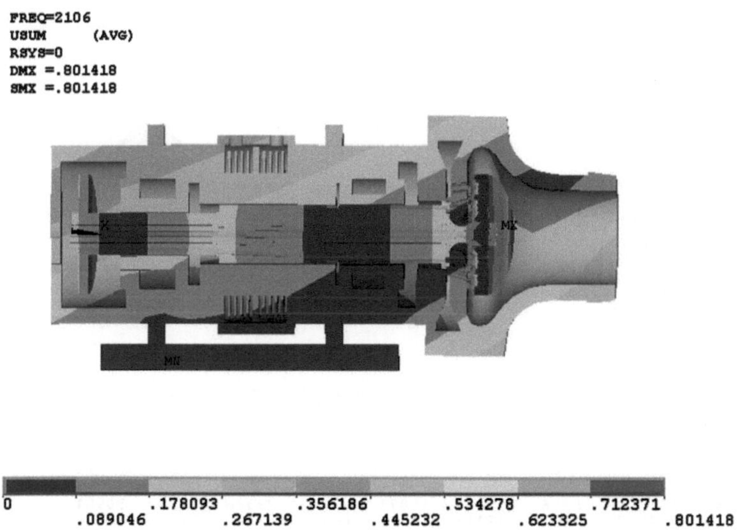

Fig. 4.34 The coupled form of vibrations of the rotor with the body at 2,106 Hz at a rotational speed of 23,800 rpm

4.1.6 Analysis of Kinetostatic Stresses

The analysis of stresses in a rotating shaft was performed on the basis of a 3D FEM model used during a computational modal analysis [7]. Figure 4.36 shows the distribution of stresses reduced on the rotor elements at the nominal speed of 23,800 rpm.

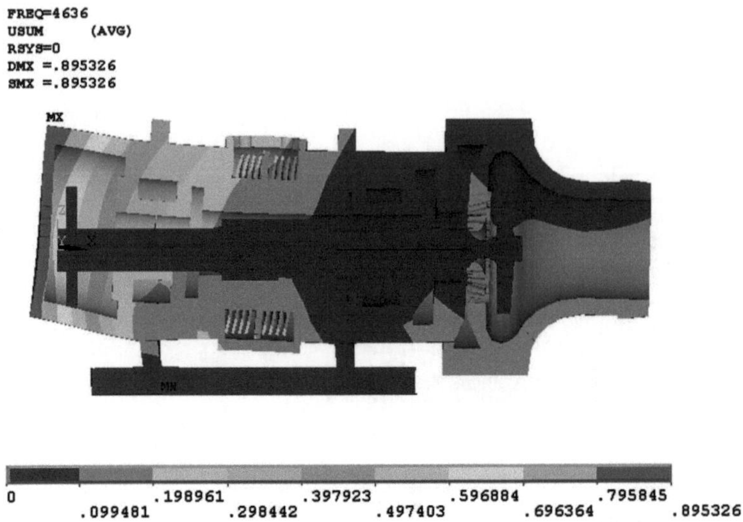

Fig. 4.35 The form of vibrations of the microturbine body at 4,636 Hz, visible clear deformation of the *left chamber* of the body

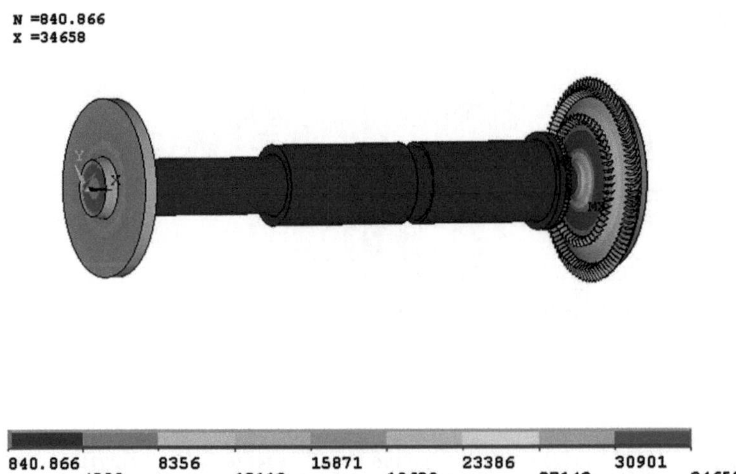

Fig. 4.36 Kinetostatic stresses in a radial microturbine rotor for the speed of 23,800 rpm

The maximum value of stresses at a speed of 23,800 rpm was 34.7 MPa. They were stresses almost ten times smaller than the acceptable stresses for the material of the rotor—steel 40H (EN 41Cr4), which amount to approx. 320 MPa. The greatest value of reduced stresses occurred in the rotor disc, at the necking of the disc (red in Fig. 4.36). Further analysis showed that at a speed of

50,000 rpm, the maximum reduced stresses amounted to 153 MPa. Despite the significant increase in the values of stresses, it can be concluded that even at this speed the stresses were acceptable. The work of the rotor at even higher speeds, however, would require another, more precise analysis of strength. This is due to the fact that the microturbine rotor disc during normal operation is affected by an increased temperature and additional aerodynamic forces associated with the flowing working medium and residual unbalance, which is the source of smaller strength and additional load.

4.1.7 Analysis of Thermal Loads

Due to the fact that high temperatures occur in microturbines, their impact on the working conditions of the machine should be assessed. Thermal loads can be triggered, among others, by asymmetrical heating of the rotor (which causes its warping), and increasing dimensions as a result of the occurrence of temperature expansion [9]. Because in the analyzed machine the rotor of the generator is an element made from a permanent magnet, one should not expect uneven heating of the shaft in the generator part. In connection with this, the focused was primarily on the phenomenon of heating of the shaft and rotor disc due to the passing thermodynamic medium [10].

The analysis was conducted for radial microturbine rotor, the construction of which has been presented earlier in Fig. 4.1. During the analysis, it was assumed that hot steam of a low-boiling medium with a temperature of 160 °C will flow onto the rotor discs [11]. In order to take into account the adverse impact of uneven heating of the rotor disc and shaft, it was assumed that steam with a temperature of 160 °C will affect half of the shield. Thermal analysis was performed using the ABAQUS package. The FEM model of a simplified rotor has been shown in Fig. 4.37.

The distribution of temperature in the microturbine rotor

In the first stage of the thermal analysis the distribution of temperature in the shaft and rotor disc was determined. The calculations assumed that the rotor disc will heat up to the temperature of the steam of the low-boiling medium. The additional

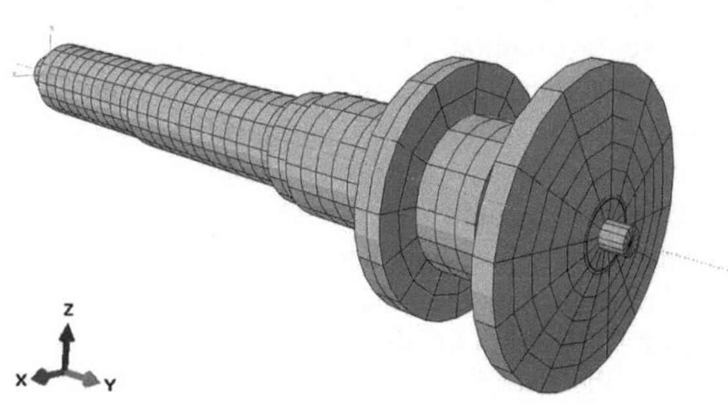

Fig. 4.37 The FEM model of the microturbine rotor used to analyze thermal phenomena

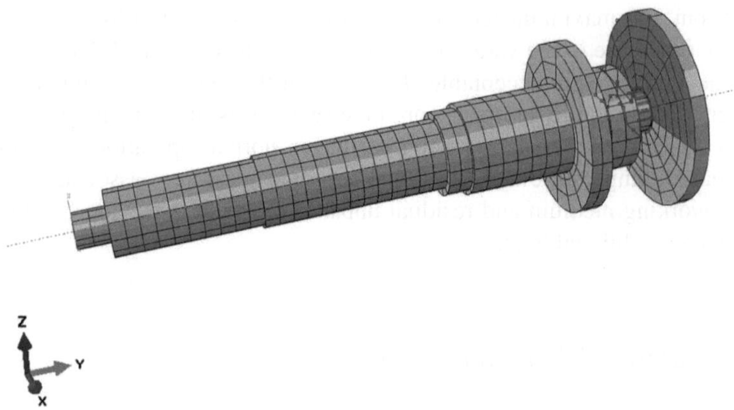

Fig. 4.38 The FEM model of the microturbine rotor with the selected surfaces at 160 °C

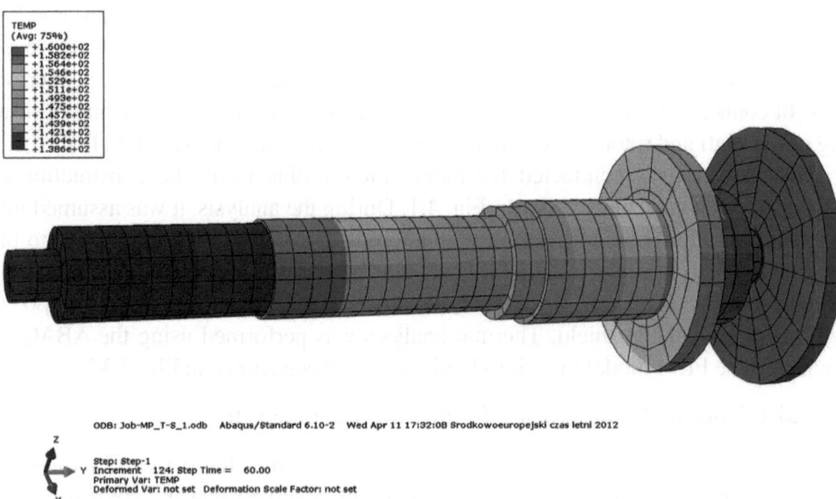

Fig. 4.39 Temperature distribution on the surface of the microturbine rotor

assumption that only half of the disc is heated to 160 °C aimed to increase the adverse distribution of temperature in the analyzed part. The heat was conducted to other elements of the rotor with a thermal conductivity coefficient of 58 W/(m K). The heat from the entire surface of the rotor expanded through radiation with the coefficient of emissivity amounting to 0.5. It was also assumed that the ambient temperature was 50 °C. The FEM model used during calculations with the marked surfaces with a temperature of 160 °C has been shown in Fig. 4.38.

The results of the calculations have been presented in Figs. 4.39 and 4.40. Figure 4.39 shows the determined distribution of temperature on the external surfaces of the rotor. The highest temperature occurred on the rotor disc and complied

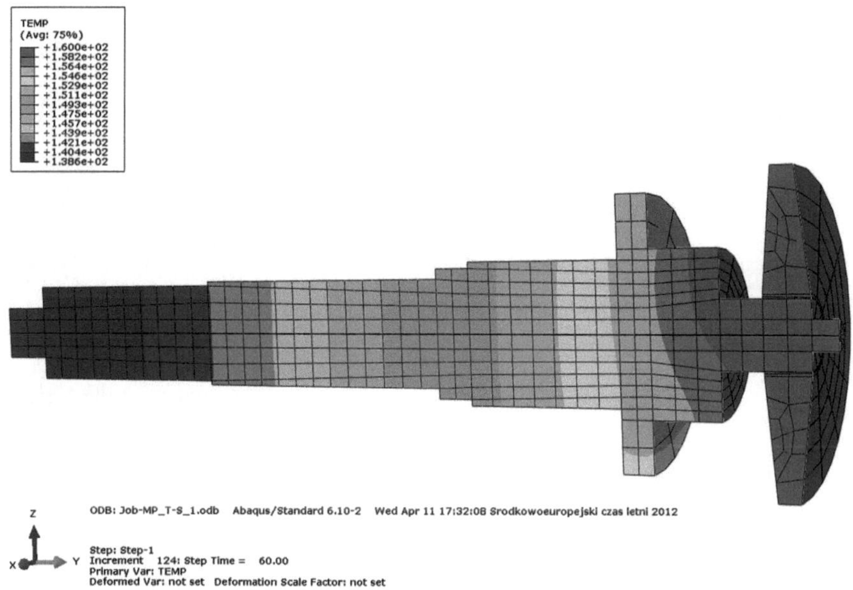

Fig. 4.40 Temperature distribution in the cross-section of the microturbine rotor

with the previously established value. In the rest of the rotor the temperature decreased as a result of the exchange of heat energy with the environment. At the other end of the rotor, due to the good thermal conductivity of steel, the temperature reached up to 139 °C (blue in Fig. 4.39). Although only the heating of half the disc was assumed, a very similar temperature occurred on its entire surface. The obtained results have been presented in Fig. 4.40 for a cross-section of the rotor. Single-sided heating of the rotor disc caused only mild disturbances of the symmetry of isotherms distribution and it took place only near the heated disc. In the rest of the shaft there were even and symmetrical temperature changes.

The diagrams presented above demonstrate that even for a disadvantageous situation, in which the passing steam of the thermodynamic factor would heat only half the rotor disc, due to the good thermal conductivity of the construction material, only small temperature gradients would occur. A worrying result, however, is the relatively high temperature of the shaft in the bearing-generator part. For a bearing journal near the fluid-flow part it amounted to as much as approx. 150 °C, in the area of the generator sleeve approx. 145 °C, and for the second bearing journal approx. 140 °C. Because gas bearings were used in the analyzed microturbine, such high temperatures should not be a source of problems during operation. Very unfavourable is the temperature of approx. 145 °C, in the area of the generator sleeve. The generator windings may in fact operate at a maximum temperature of 150 °C. In view of the fact that its temperature rises anyway in the process of generating electricity, it should be noted that, in order to ensure its correct operation, it is necessary to use additional external cooling, which was provided in the construction of the microturbine body (Fig. 4.1).

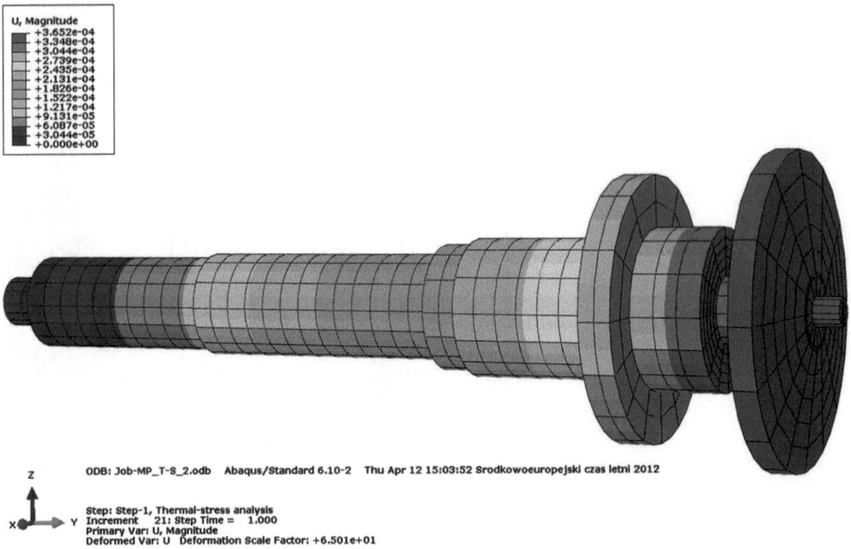

Fig. 4.41 The distribution of displacements of the microturbine rotor caused by heating during operation

Distribution of deformations and thermal stresses

On the basis of determined temperature distributions, a structural analysis was conducted, which aimed to determine the displacements, deformations and thermal stresses. During the analysis, the capabilities of the ABAQUS system were used in the scope of coupled analyses, enabling to include the results of thermal analysis in the structural calculation. Temperature distributions discussed in the previous section have been included in the structural analysis as load. The ambient temperature, which was also the initial temperature of the rotor, was assumed at 20 °C. This means that the thermal increase of dimensions of the rotor was determined by its temperature.

In order to carry out a structural analysis the following boundary conditions were adopted. The displacements of the free end of the rotor (without a disc) were blocked in all directions through the reaffirmation of the node located on the axis of rotation. Additionally, at the same end, the free rotation of the rotor was blocked. The other end of the rotor, near the rotor disc, could move only in the axial direction. Such conditions of anchoring allowed to determine displacements, deformations and stresses caused by temperature changes, without affecting the external anchoring.

The determined distribution of displacements for the microturbine rotor has been presented in Figs. 4.41 and 4.42. As a result of temperature expansion the entire length of the rotor increased by 0.37 mm. Displacements only in the axial direction have been presented in Fig. 4.43. The increase in the diameter of the rotor disc, however, was 0.13 mm.

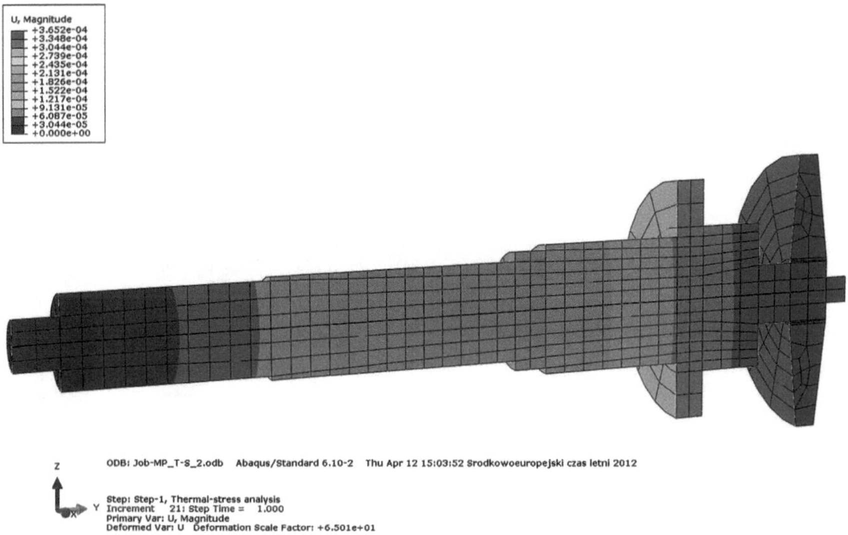

Fig. 4.42 The distribution of displacements in the cross-sections of the microturbine rotor caused by heating during operation

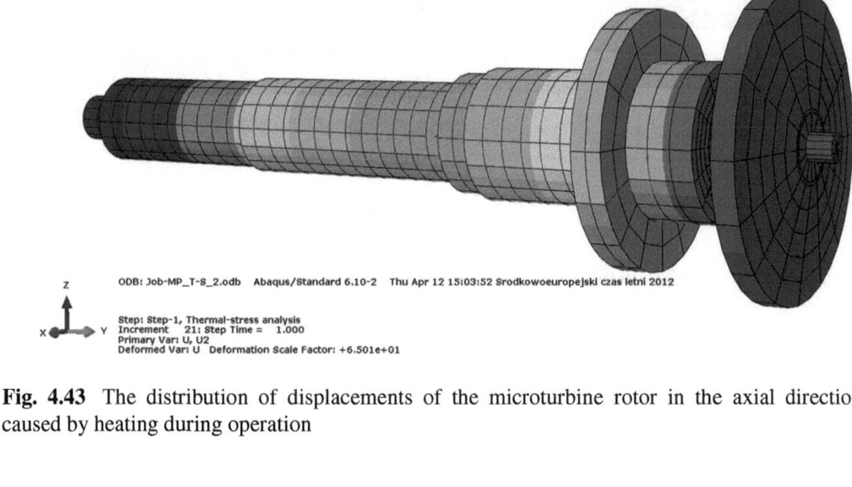

Fig. 4.43 The distribution of displacements of the microturbine rotor in the axial direction caused by heating during operation

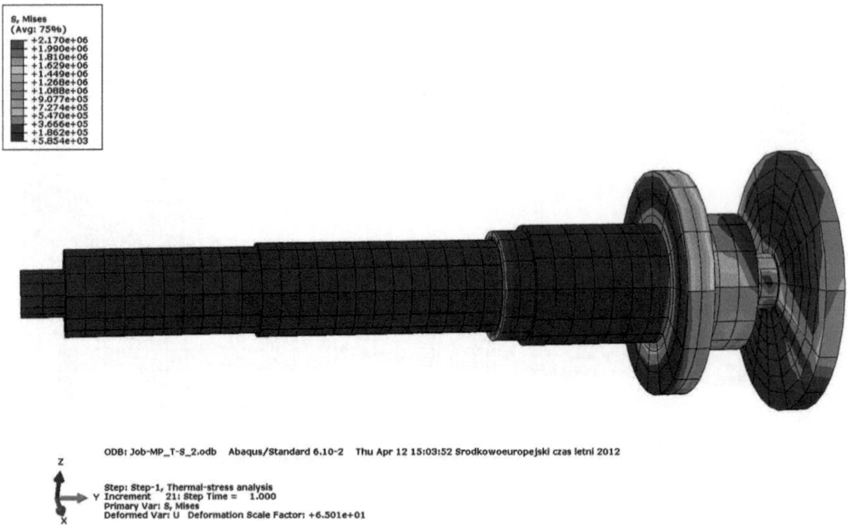

Fig. 4.44 The distribution of thermal stresses in the microturbine rotor

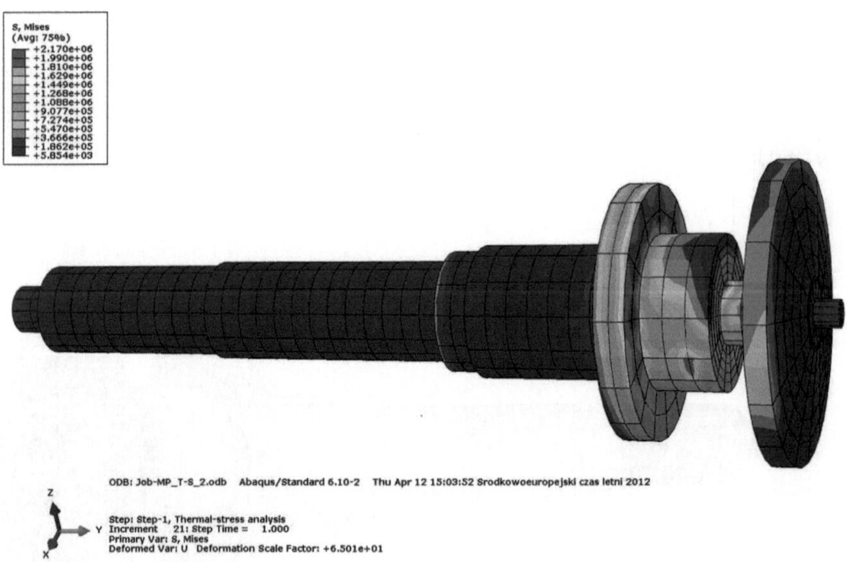

Fig. 4.45 The distribution of thermal stresses in the microturbine rotor

The following drawings (Figs. 4.44, 4.45 and 4.46) present the distributions of thermal stresses. The maximum thermal stresses reached up to 2.2 MPa and have been marked on the drawings below in red. They occurred in the disc between the seal and the bearing and at the location of the sudden change in the diameter of the shaft (Figs. 4.45 and 4.46).

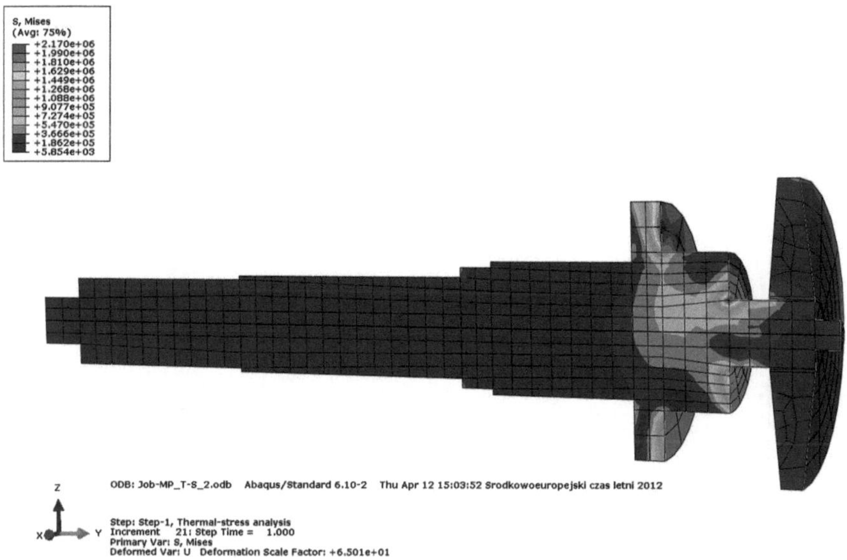

Fig. 4.46 The distribution of thermal stresses in the cross-section of the microturbine rotor

To summarize the analysis of thermal loads of the microturbine rotor it can be concluded that during normal operation there should not be any problems with thermal stresses. This is due to the fact that thanks to the good conductivity of the construction material in the entire rotor, there was a uniform distribution of temperature. Because even at unequal heating of the rotor no significant warping of the shaft was observed, the impact of any thermal warping on the dynamics of the rotating system was not analyzed. The biggest thermal threat for a tested rotor seems to be its high temperature in the generator part. The obtained results show that in order to ensure the proper functioning of the generator, it will be necessary to use external cooling provided in the construction. Since the analysis carried out also revealed a rather large extension of the shaft, it is necessary to take it into account in the selection of axial clearances in longitudinal bearings, and also provide in the start-up procedure of the microturbine the need to heat the body evenly.

4.1.8 Analysis of Stiffness of Bearing Supports

The stiffness of bearing supports can have an impact on the critical speed of the rotating system and the amplitude of vibrations in the entire range of rotational speed [9]. The characteristics of rotating systems determined at the assumption of ideal stiffness of supports most often do not coincide with the results of experimental studies. This is one of the most important causes of non-compliance of theoretical and experimental characteristics.

Table 4.4 The stiffness and damping coefficients of aerodynamic gas bearings adopted for calculations

No. of bearing	Stiffness coefficient (N/m)		Damping coefficient (N s/m)	
	γ_{22}	γ_{33}	β_{22}	β_{33}
1	8.3×10^6	8.5×10^6	5×10^3	5×10^3
2	8.3×10^6	8.4×10^6	5×10^3	5×10^3

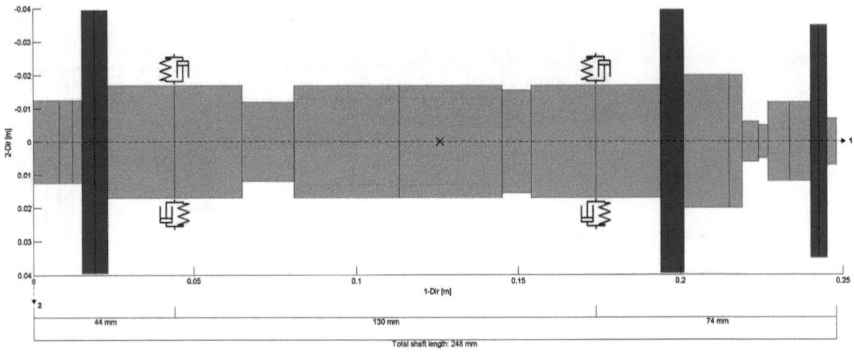

Fig. 4.47 The rotor model of the microturbine developed using the program MADYN 2000

As part of the ongoing analyses, the impact of the stiffness of bearing supports on the dynamic properties of the microturbine rotor was tested, assuming that it has aerodynamic gas bearings. The determined stiffness and damping coefficients of the bearings, which were used during the analysis have been provided in Table 4.4. The analysis was performed using the program MADYN 2000, assuming constant characteristics of bearings. The rotor model has been shown in Fig. 4.47.

Rigid bearing supports

The base characteristics, which were used for comparison, were the courses obtained assuming an infinitely high stiffness of bearing supports. As a reference case the variant was selected in which the unbalance of the rotor disc occurred (disc no. 4) with a value of 1.2×10^{-6} kg m. A graphic presentation of this variant has been provided in Fig. 4.48. The courses of vibration amplitudes for bearings and 3 discs (located on two ends and between the bearings) have been presented in Fig. 4.49. In order to be able to observe critical speed of the rotor in the presented range of rotational speeds, the scope of analyses was extended to 150,000 rpm.

The presented characteristics show that the critical speed of the tested system occurred only at approx. 130,000 rpm, which corresponds to the frequency of the second form of bending vibrations of the rotor presented earlier in Fig. 4.32. At that speed the vibration amplitudes of discs and journals of bearings reached

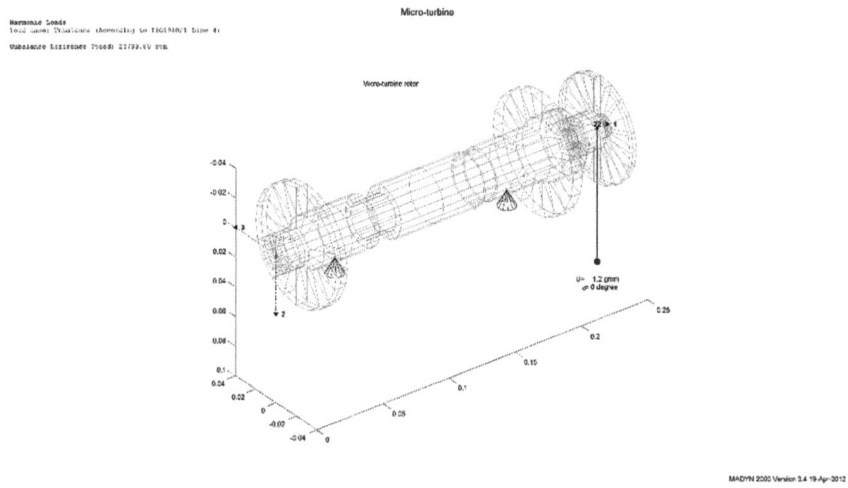

Fig. 4.48 The rotor model of the microturbine with unbalanced disc no. 4

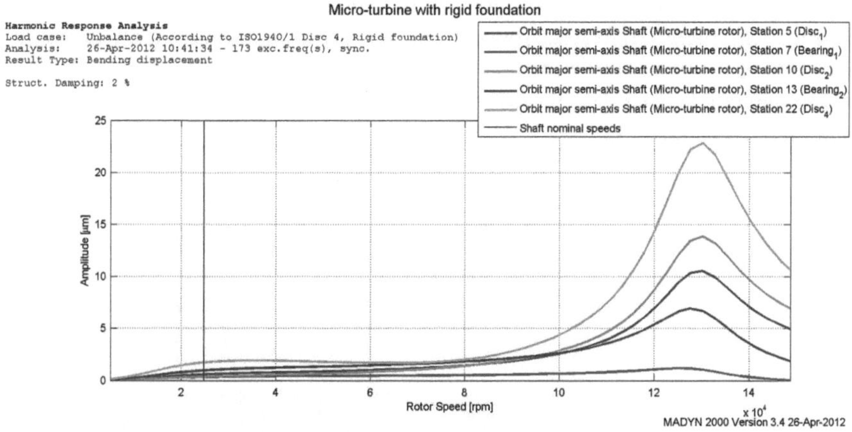

Fig. 4.49 The course of vibration amplitudes of the microturbine rotor with a rigid support of bearings

maximum values. The largest amplitude of vibrations reaching up to 24 μm occurred for the rotor disc, found at the free end of the shaft. On this shield unbalance was modelled. In terms of lower speeds (pre-resonance) the system worked very steadily and the vibration amplitudes did not exceed 2 μm. At the resonance speed the vibration amplitude in the bearings reached more than a few μm, which meant they worked steadily. After passing through the resonance area, another drop in the level of vibrations could be observed.

In order to assess the impact of the sensitivity of bearing supports on the dynamics of the rotor, the characteristics discussed in this section, was then compared with the results obtained for the system taking into account the stiffness characteristics of the supporting system.

Sensitive bearing supports

In order to determine the stiffnesses of bearing supports of the radial microturbine, additional analyses were conducted using ABAQUS software. Tests of the stiffness of bearing supports were done on the basis of a 3D FEM model of the microturbine body. This model, in relation to the actual construction, was partly simplified. The parts which did not have a significant impact on the stiffness of bearing supports were omitted. The model used has been shown in Fig. 4.50.

On the basis of the presented model the static stiffness of bearing supports was determined. The analysis consisted in driving displacements of bearing supports with force acting in the plane perpendicular to the axis of the rotor horizontally and vertically. The amplitude of the driving force amounted to 20 N and was estimated as the sum of the maximum static and dynamic strength, which can be transferred by the bearing at the nominal speed. Examples of the results of the analysis in the form of the distribution of displacements on the microturbine body caused by the activity in support no. 1 of the horizontal force have been presented in Fig. 4.51. The test results showed that the greatest displacements of bearing supports at a force of 20 N reached up to 18.2 μm (in support no. 2 in the horizontal direction). The determined static stiffnesses of bearing supports have been summarized in Table 4.5.

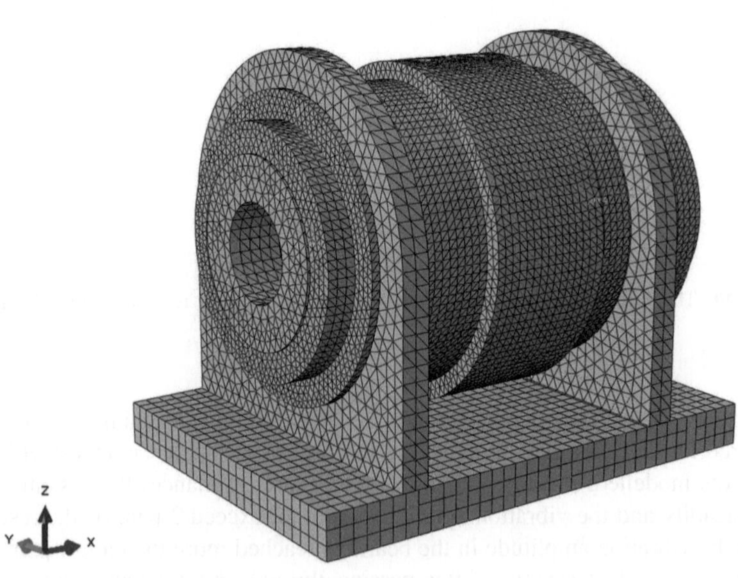

Fig. 4.50 The simplified FEM model of the body of a radial microturbine developed for the analysis of bearing supports stiffness

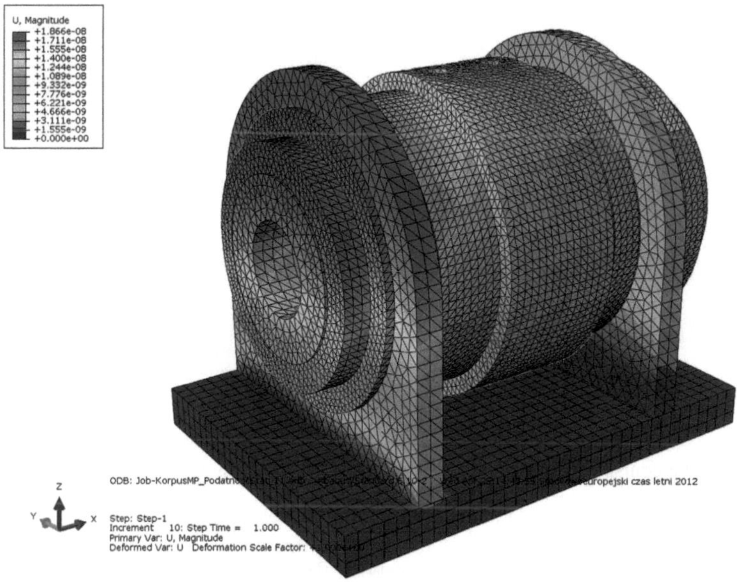

Fig. 4.51 The distribution of displacements in the microturbine body caused by static force acting on bearing support no. 1 in the horizontal direction

Table 4.5 Static stiffness of bearing supports

	Bearing support no. 1		Bearing support no. 2	
	Displacement (μm)	Stiffness (N/m)	Displacement (μm)	Stiffness (N/m)
Horizontal direction	13.4	1.5×10^6	18.2	1.1×10^6
Vertical direction	4.6	4.3×10^6	7.4	2.7×10^6

The precise determination of stiffnesses of bearing supports would require the performance of dynamic tests (experimental or simulation) with regard to the frequency of the driving force. The procedure for determining the dynamic stiffness is quite complex and has been described in detail e.g. in works [9, 12, 13]. Such an analysis should be begun by determining the dynamic sensitivity of bearing supports, and then on its basis the sought stiffness can be determined based on one of the methods of transformation of dynamic characteristics. Therefore, it is necessary to carry out time-consuming experiments (or calculations), and apply a complex mathematical apparatus. Due to the auxiliary nature of the performed analyses and basing on the model of only the body of the micro power plant, without taking into account how it is set in the target micro power plant, more advanced and time-consuming analyses of dynamic stiffness of bearing supports would be unjustified in this case. Additionally, in this case using a simplified method is advocated by the fact that the method itself of mounting the microturbine in the micro power plant can be crucial for the stiffness of the supporting system.

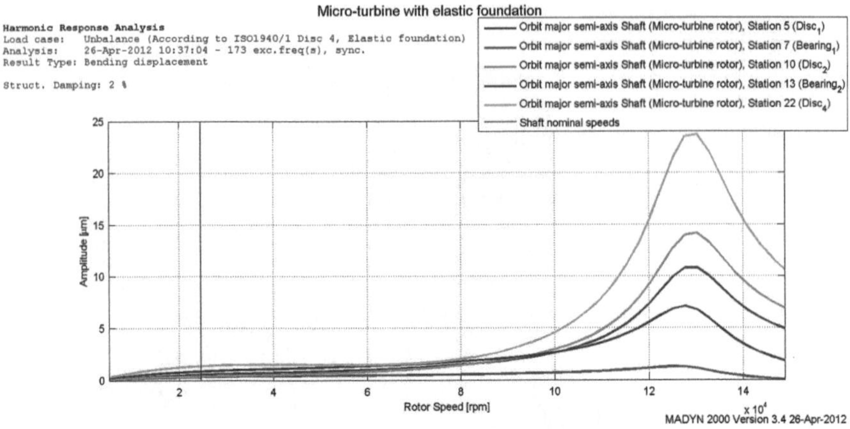

Fig. 4.52 The course of vibration amplitudes of the microturbine rotor including the stiffness of bearing supports

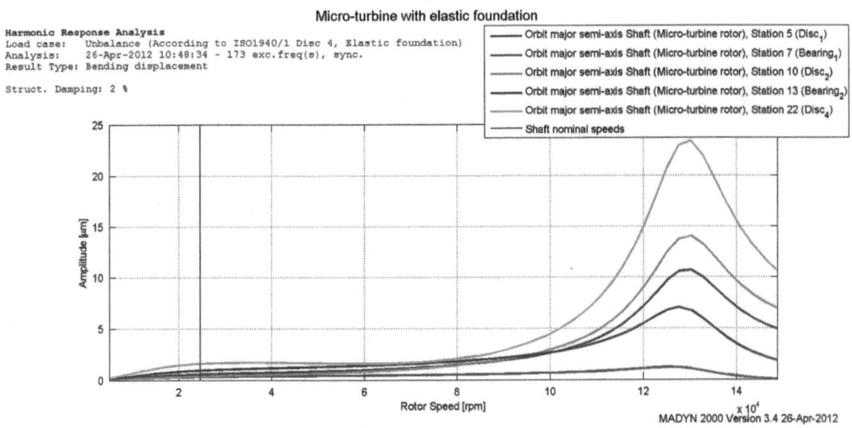

Fig. 4.53 The course of vibration amplitudes of the microturbine rotor including the stiffness of bearing supports (stiffness 5 times greater than the base one)

In light of the above, further analysis was performed on the basis of the determined static stiffness. It was also assumed that the stiffness of both supports was the same and amounted to 1.3×10^6 N/m in the horizontal direction and 3.5×10^6 N/m in the vertical direction. In carrying out a parametric analysis, the determined values were treated as base values, by also analyzing cases with a stiffness that was 5 and 10 times greater. These values were placed into the numerical model developed in the program MADYN 2000. The obtained results have been presented in Figs. 4.52, 4.53 and 4.54.

The obtained results of the calculations in the first assessments proved to be quite surprising. For a system in which the sensitivity of bearing supports was taken into account, almost identical results were obtained as before—with a rigid support of

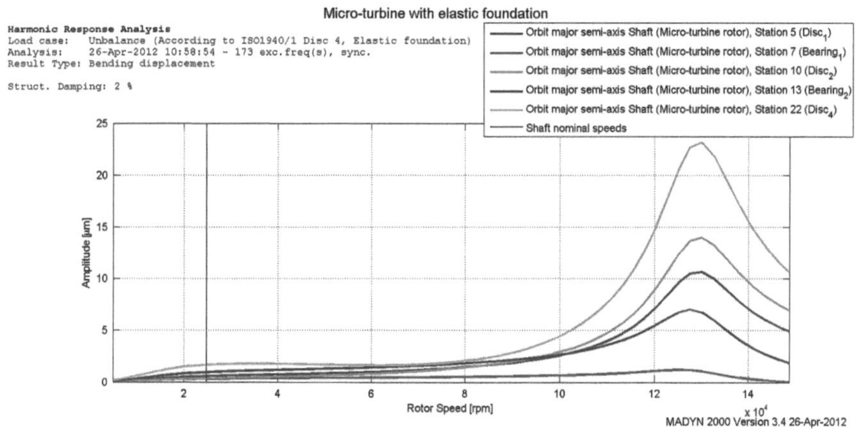

Fig. 4.54 The course of vibration amplitudes of the microturbine rotor including the stiffness of bearing supports (stiffness 10 times greater than the base one)

bearings. The subtle differences were noticed only in respect to the lower rotational speeds (within the nominal speed). A detailed analysis of the studied cases showed, however, that changes in the stiffness of bearing supports affect the substitute stiffness in points of support of the rotor to a small extent. The substitute stiffness is determined on the basis of stiffnesses of bearings and bearing supports [9, 12, 13]. It turned out that due to the relatively low stiffness of aerodynamic gas bearings, even drastic changes in the stiffness of the bearing support did not have a great impact on the dynamics of the rotor. With high stiffness of the bearing support, the substitute support had a similar value to the stiffness of the bearings themselves. With low stiffness of bearing supports, the substitute stiffness decreased, however, this did not affect the high-frequency vibrations of the rotor, which was connected to its very high stiffness (large diameter at a small length). The only sign of a change in the stiffness of the supports were the already previously mentioned changes in vibration amplitudes at speeds close to the nominal speed. At the lowest stiffness of bearing supports the vibration amplitude of disc no. 4 decreased by approx. 1 μm, which from a practical point of view did not make any difference. These changes resulted from the change in the modal characteristics of the system, namely to the change in the frequency of occurring vibrations of a stiff rotor in the bearings (cylindrical and conical vibrations). Due to the fact that these forms of vibrations are strongly dampened, they have very little effect on the dynamics of the rotor.

4.2 Axial Microturbine

The axial microturbine designed by the team of prof. Kosowski [14–17] from the Gdańsk University of Technology is a five-stage turbine with partial supply. In this solution, all the turbine stages consisted of rotor discs of the same geometry, and

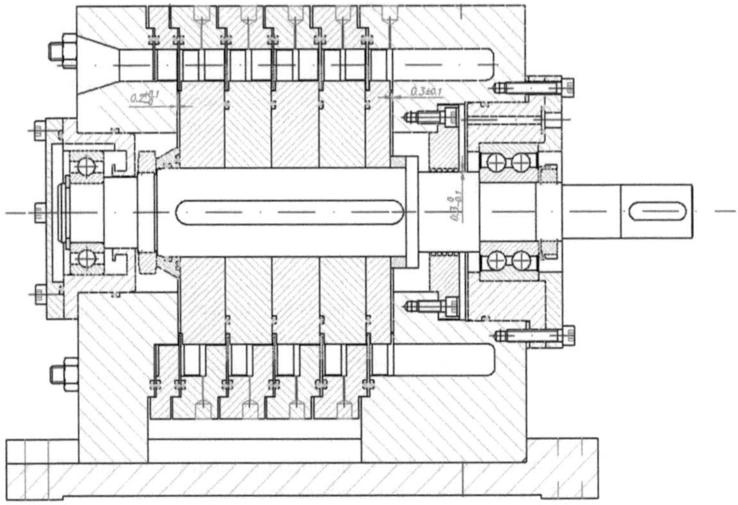

Fig. 4.55 Cross-section of an axial microturbine [17]

each discs had 71 identical blades. The diverse flow between different turbine stages was obtained by changing the number of guide blades (variable radius of power supply). The structural solution of this variant of the microturbine has been shown in Fig. 4.55. The dividing diameter of the rotor discs was approx. 100 mm, and the overall length of the rotor—230 mm. The theoretical power of the turbine was determined at the level of 3 kW at 8,073 rpm [15, 17]. To work with the designed microturbine, a three-phase synchronous generator was foreseen, connected with the microturbine shaft by a clutch. The maximum power of the generator used amounted to 4 kW. During initial start-up tests on the test bench, rolling bearings were used, which are easily accessible and at the stage of initial tests do not need to have features such as high durability or low resistance to motion. Ultimately, in this microturbine it is planned to apply rolling bearings made from ceramic materials.

An unquestionable advantage of this design is its simplicity, which consists in applying a few discs with blades of the same geometry. This significantly facilitates the creation of blade discs and later exchange of used or damaged parts. The use of the same geometry of blades in subsequent stages, however, takes place at the expense of the efficiency of the fluid-flow system. A certain problem in this microturbine may also be the relatively great axial forces occurring during operation, which must be transferred by the bearings.

4.2.1 Analysis of Rotor with Slide Bearings

Within the framework of the simulation tests, the possibility of using axial slide bearings lubricated by a low-boiling liquid medium was examined. In contrast to rolling bearings, slide bearings could provide quiet and stable operation of the

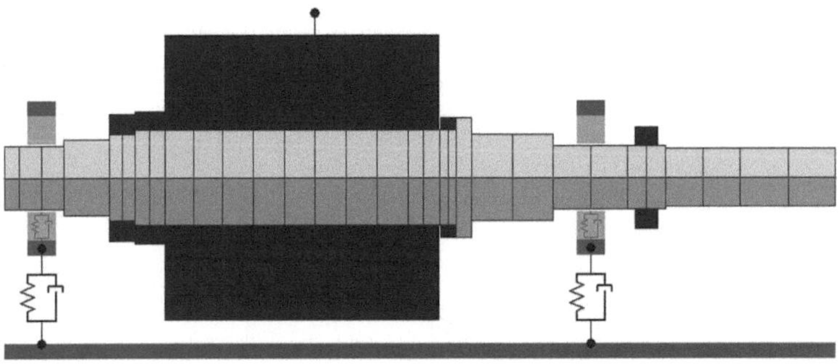

Fig. 4.56 The FEM model of the axial microturbine rotor with slide bearings developed in the system MESWIR

rotor, and the use of the same mixture as the lubricant medium for powering the microturbine would significantly simplify the construction of the machine.

The FEM model of a five-stage axial turbine used during the analysis of transverse vibrations of the rotor-slide bearings system has been shown in Fig. 4.56. It was developed in programs from the MESWIR series. Because for this fluid-flow machine the nominal speed amounted to 8,073 rpm, the calculations were carried out in the rotational speed range from 1,000 to 10,000 rpm. The selected, as a result of the analysis, optimal dimensions of bearings for this rotor (due to the stability of the system) were as follows:

- Bearing 1: diameter of journal—20 mm, width of sleeve—8 mm, radial clearance—12 μm;
- Bearing 2: diameter of journal—20 mm, width of sleeve—8 mm, radial clearance—12 μm.

It should be emphasized that the diameters of bearing journals, as in the case of the radial microturbine, were not changed, because they were imposed by the constructors of the machine. However, the width of the bearing and size of the oil clearance were optimized. The course of relative vibrations amplitudes journal-sleeve, the shape of the rotating shaft line and diagnostic card for speeds equal to 8,000 rpm have been presented in Figs. 4.57, 4.58 and 4.59.

The studies showed that after a suitable selection of the geometry of bearings, it was possible to obtain stable operation of the rotor in the entire analyzed range of rotational speeds. Along with the increase in the rotational speed of the vibrations amplitude in bearings clearly increased, however, even at a speed of 10,000 rpm they did not exceed 3 μm (Fig. 4.57). Stable operation of the system is also visible on the diagnostic card presented in Fig. 4.59, where a clear dominance can be seen of the stripe corresponding to the first harmonic component. This component had a dominant character also at other rotational speeds.

The results of the calculations obtained indicate that the rotor of the axial microturbine could be supported in two hydrodynamic slide bearings lubricated by a low-boiling medium in liquid form. Such a bearing system would ensure stable operation of the microturbine.

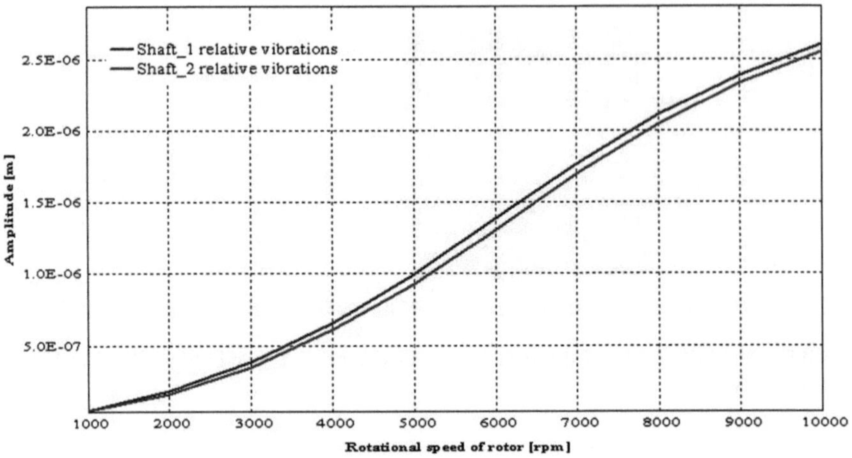

Fig. 4.57 The course of amplitudes of journal-sleeve relative vibrations for the axial microturbine in the scope of rotational speeds from 1,000 to 10,000 rpm

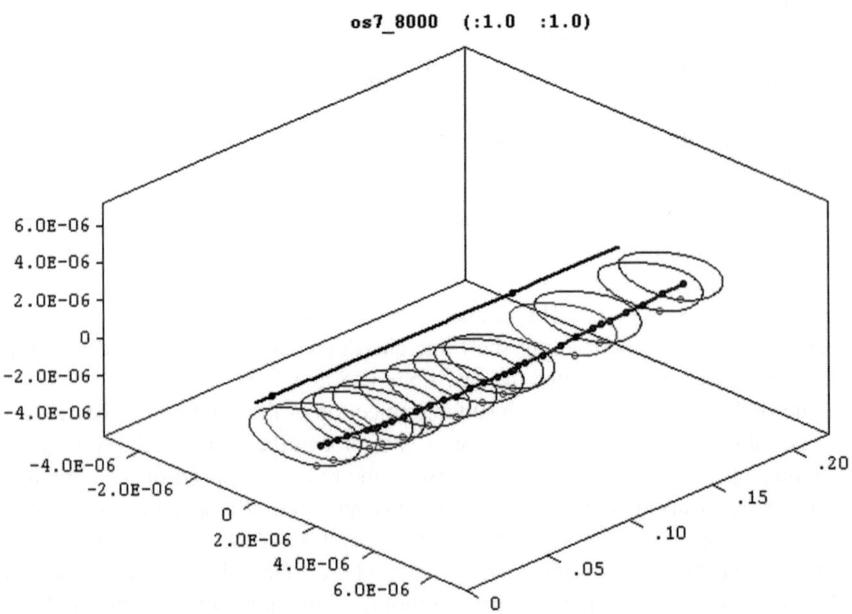

Fig. 4.58 The shape of the rotating rotor line of the axial microturbine with slide bearings at a speed equal to 8,000 rpm

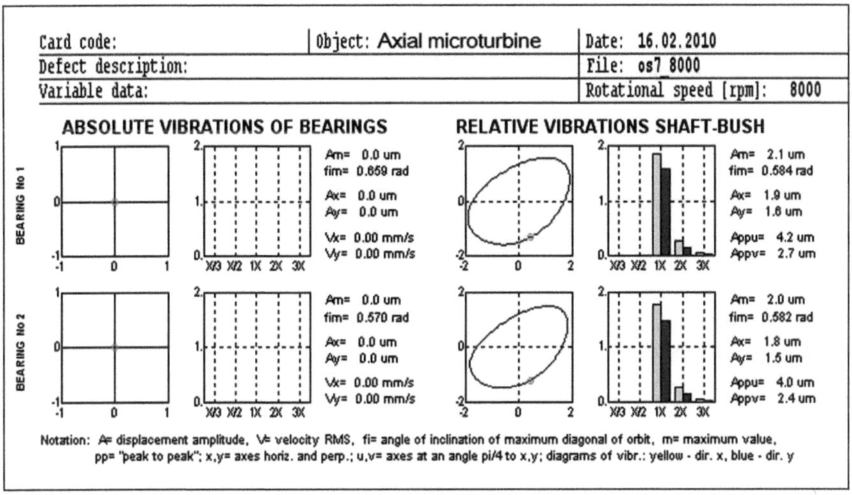

Fig. 4.59 Diagnostic card for the axial microturbine with slide bearings at a speed equal to 8,000 rpm

4.2.2 Analysis of Rotor with Rolling Bearings

The discussed studies of the axial microturbine rotor with rolling bearings were carried out using the program MADYN 2000. The calculations included the simplified scheme of loads acting on the rotor during normal operation [18]. The values of axial forces and moments acting on individual stages have been provided in Table 4.6. It was assumed that the entire torsional moment is transferred through the inlet connection located at one end of the shaft. During the analysis of forced vibrations of the analysis, the unbalance of the rotor was modelled using the standard ISO 1940/1. The class of unbalance G 2.5, suitable for gas and steam turbines was assumed.

In this microturbine, the constructors suggested the use of the following rolling bearings: transverse ball bearing, single-row 6,004, and diagonal ball bearing, double-row 3,204. For these bearings the substitute stiffnesses were determined using the theoretical dependencies provided in the work [19]. The determined values have been provided in Table 4.7.

The provided loads and substitute stiffness coefficients of rolling bearings were implemented into the numerical model of the rotor. The scheme of the axial microturbine rotor divided into finite elements has been shown in Fig. 4.60. The model consisted of 31 beam elements. In the diagram, the discs made of aluminium have been marked by a darker colour. The nominal rotational speed was adopted as for an actual microturbine at 8,073 rpm.

In the first stage of calculations, a kinetostatic analysis was performed, within which displacements of journals were determined, as well as deformations of the

Table 4.6 The values of forces and moments adopted for calculations

No. of microturbine stage	Axial force (N)	Torsional moment (Nm)
1	10.6	0.92
2	17.7	0.82
3	9.4	0.83
4	4.3	0.90
5	1.2	1.44

Table 4.7 Determined stiffness coefficients of rolling bearings

Bearing label	6004 (N/m)	3204 (N/m)
Radial stiffness	3.14×10^7	2.69×10^7
Axial stiffness	–	2.83×10^7

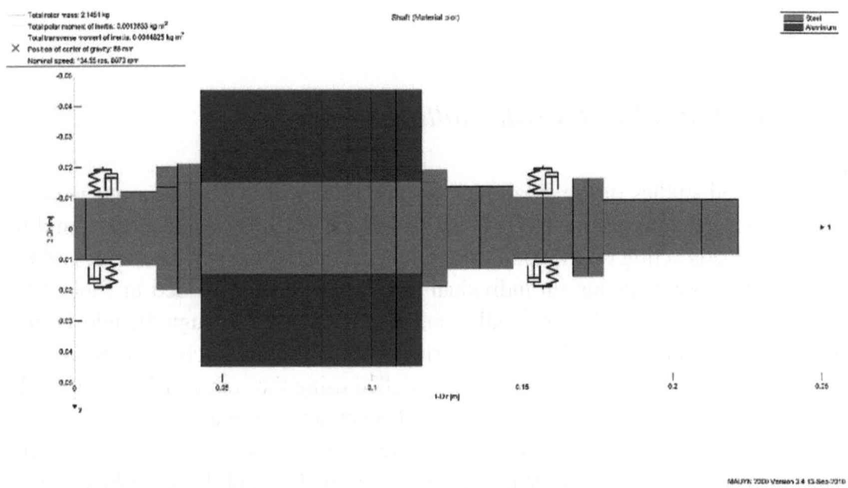

Fig. 4.60 The model of the axial microturbine rotor with rolling bearings and discs made from a different material

shaft caused by gravity. The results of these calculations have been presented in Fig. 4.61. Due to the fact that the rotor was characterized by small dimensions and low weight, and the bearings with high stiffness, the displacements caused by gravity amounted to a maximum of 0.4 μm at the free end of the shaft.

The next stage of the analysis of the rotor-bearings system was a dynamic analysis of forced vibrations. The unbalance of the rotor was applied to the third, middle disc. The studies of forced vibrations were carried out at the speed range from 1,000 to 30,000 rpm. As a result of the calculations, courses of vibrations

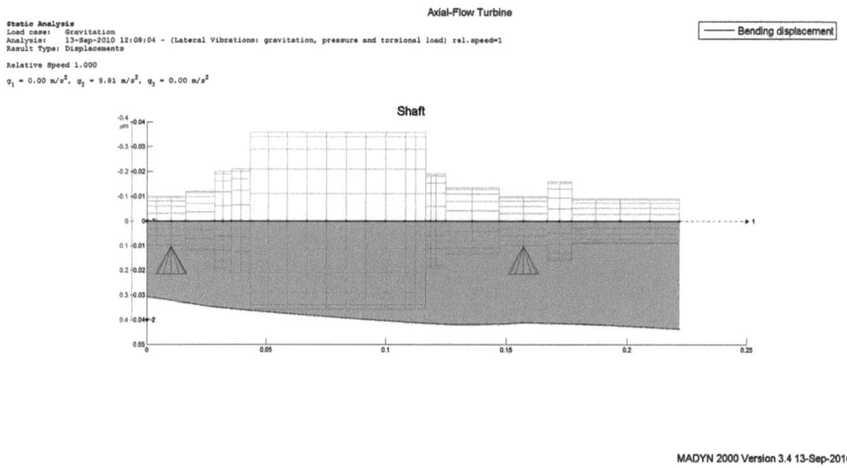

Fig. 4.61 Transverse displacements of the rotor caused by gravity

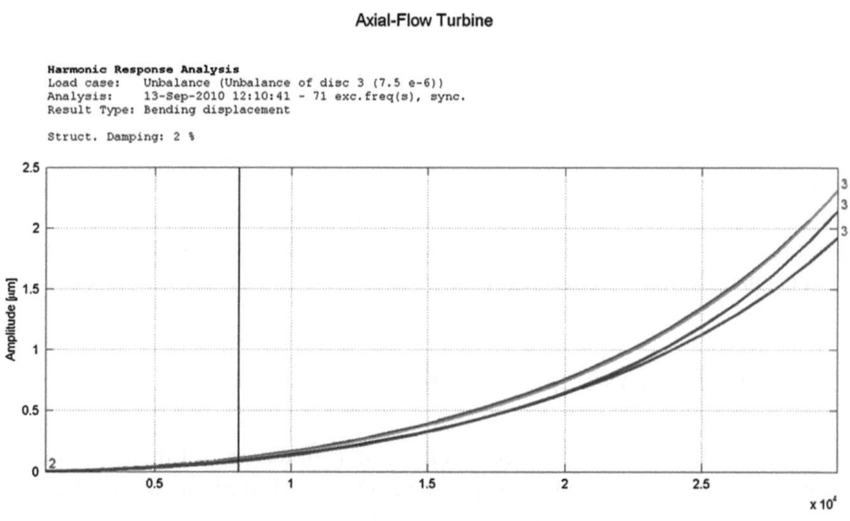

Fig. 4.62 The characteristics of dynamic responses of the rotor-bearings system of the axial microturbine

amplitudes of chosen nodes of the system were obtained, which has been presented in Fig. 4.62. The results presented concern: two bearing journals, the middle disc and free end of the shaft. The trajectories of vibrations had a similar shape of a circle, which has been presented in Fig. 4.63.

In general, it can be concluded that in the considered range of speeds, very small vibrations amplitudes occurred not exceeding 2.5 μm. Because the tested system

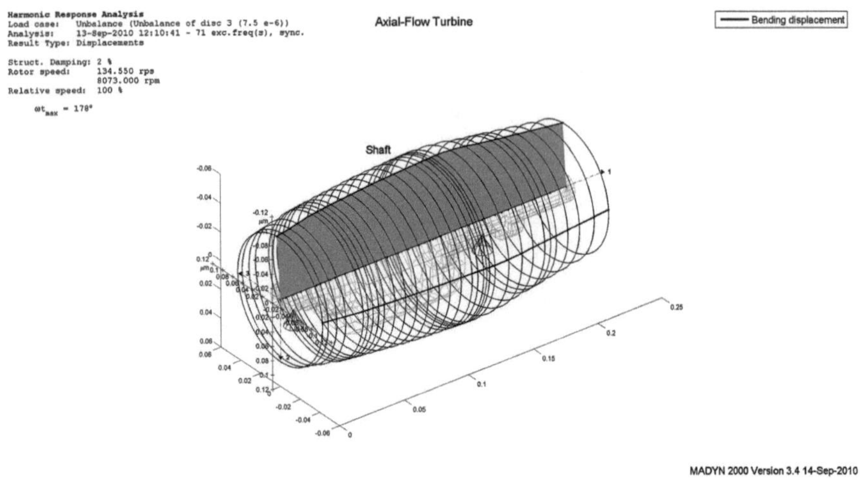

Fig. 4.63 The line shape of the rotor rotating at a speed of 8,073 rpm

was subcritical, the values of vibrations amplitudes steadily increased along with the increasing rotational speed. In the analyzed speed range there was no resonant speed.

Within the framework of the analysis of the axial microturbine with rolling bearings, a strength analysis of the rotor was also carried out. The values of stresses occurring in the rotor elements were significantly lower than the acceptable stresses of the construction materials. Detailed results of this analysis have been presented in the work [18]. The cited work also discusses the analysis of longitudinal and torsional vibrations of the rotor.

4.2.3 Modal Analysis of the Rotor

Modal tests of the axial microturbine rotor, just as in the case of the radial microturbine, were performed using the program ANSYS. The 3D model of the axial microturbine rotor used during modal simulation analysis has been presented in Fig. 4.64. For the construction of the mesh of finite elements, isoparametric eight-node elements were used. The entire model had more than 380 thousand degrees of freedom. The bearings were modelled by elastic elements [7].

The analysis of free vibrations was carried out for a rotor not rotating and rotating at a nominal speed of 8,073 rpm. The values of indicated frequencies of free vibrations have been provided in Table 4.8.

The first six forms of vibrations for the model of the axial turbine have been shown in Table 4.9. The first series of vibrations of the rotor shield began from the frequency 12,339 Hz, at which there were vibrations of all the blades (Table 4.9).

Fig. 4.64 The 3D model of the axial microturbine rotor

Table 4.8 The frequencies of the axial microturbine rotor's free vibrations not rotating and rotating at a speed of 8,073 rpm

No.	0 rpm	8,073 rpm
	Frequency	
	(Hz)	(Hz)
1	2,487.6	2,488.0
2	2,487.6	2,488.0
3	8,583.4	8,583.5
4	8,583.4	8,583.5
5	10,636	10,636
6	11,968	11,969
7	11,968	11,969
8	12,339	12,343
9	13,163	13,168
10	13,163	13,168
11	13,176	13,181
12	13,176	13,181
13	13,261	13,266
14	13,261	13,266
15	13,281	13,294

These results were obtained for the stiffness coefficients of bearings appropriate for hydrodynamic slide bearings lubricated by a low-boiling liquid medium [4]. The modal analysis of the rotor was also carried out taking into account slide bearings, obtaining similar results [18]. The first form of deflective vibrations of the shaft occurred at a slightly higher frequency (which was due to the greater stiffness of bearings), and also two forms of vibrations were obtained for the rotor as a rigid solid. Regardless of the adopted bearing system, the tested rotor was subcritical, as the nominal rotational frequency was significantly lower than the frequency of the first form of deflective vibrations.

4.2.4 Modal Analysis of the Microturbine

The 3D model of the axial microturbine rotor in the body has been presented in Fig. 4.65. The FEM model was developed on the basis of this model. Discretization of the microturbine body has been presented in Fig. 4.66. The modal calculations were carried out using ANSYS software. The combination of the rotor with the body was modelled using elastic elements of a stiffness indicated for hydrodynamic slide bearings, just like in the previous section. The tests mainly aimed to determine the frequencies of free vibrations of the body and reveal any couplings between the vibrations of the rotor and the body [20].

The simulation tests conducted showed that the couplings of vibrations of the rotor and body could only occur in the case of a high stiffness of connection between these two parts (much greater than it can take place in the case of bearings). Additionally, it was noticed that due to the sensitivity of the body, the first form of deflective vibrations of the rotor occurred at a lower frequency than was the case in the analysis of only the rotor (Fig. 4.67). The most interesting results

Table 4.9 The initial forms of vibrations of the axial turbine rotor

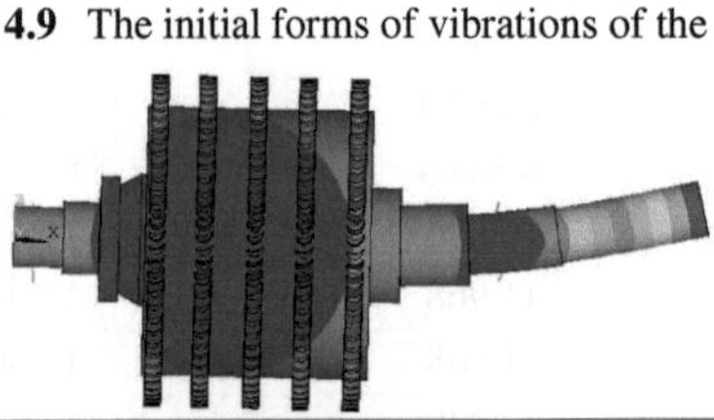	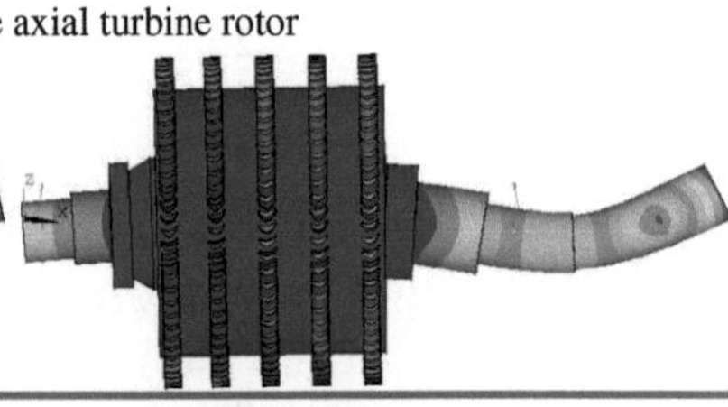
The first deflective form of vibrations of the rotor at a frequency of 2487.6 Hz	The second deflective form of the rotor at a frequency of 8583.4Hz
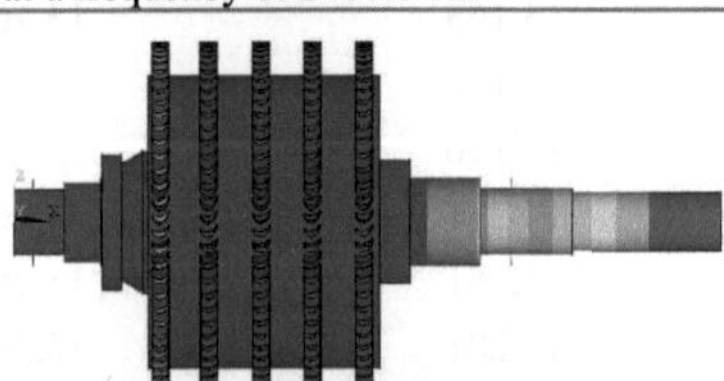	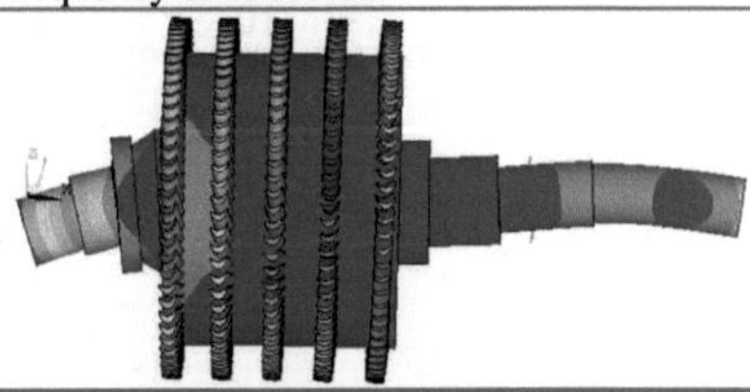
The first torsional form of the rotor at a frequency of 10636 Hz	The deflective form at a frequency of 11968Hz
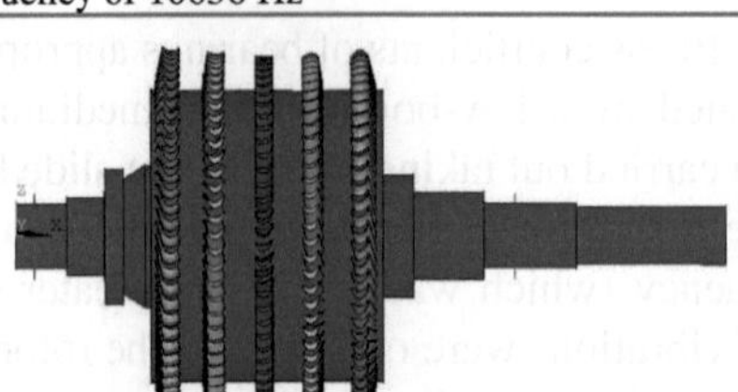	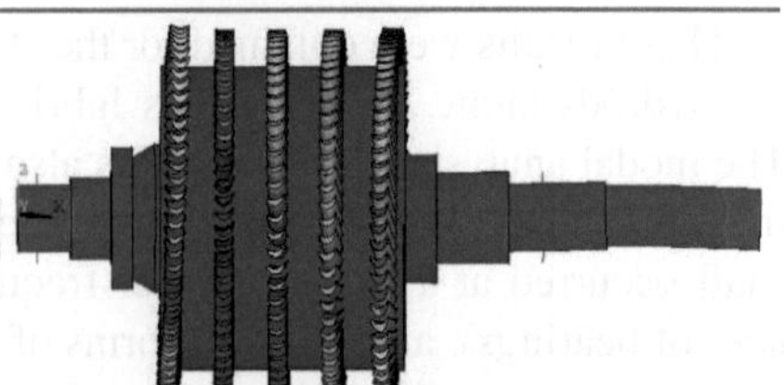
Zero node diameter of blade discs at 12339 Hz	One node diameter of blade discs at 13163 Hz

Fig. 4.65 The 3D model of the rotor and body of the axial microturbine

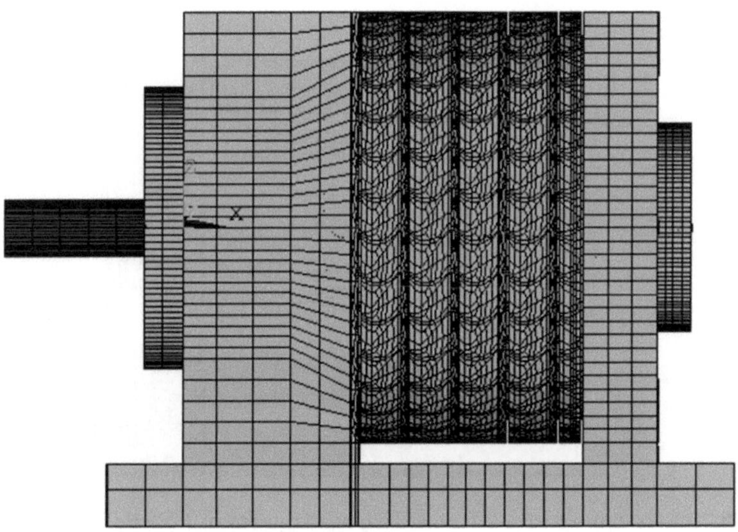

Fig. 4.66 The FEM model of the axial microturbine

of these calculations have been shown in the following diagrams (Figs. 4.67, 4.68, 4.69 and 4.70). Figure 4.68 shows the vibrations of the body at a frequency of 2,225 Hz. After applying a rigid connection between the microturbine subassemblies, the couplings of vibrations presented in Figs. 4.69 and 4.70 were obtained.

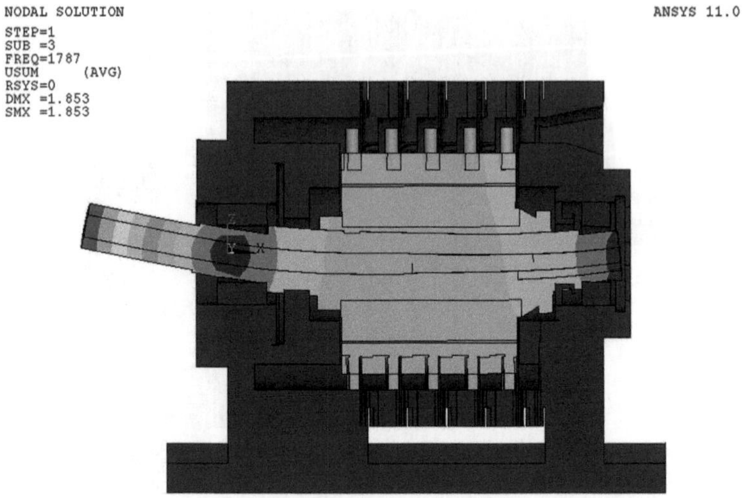

Fig. 4.67 The bending form of the rotor in the body at a frequency of 1,787 Hz

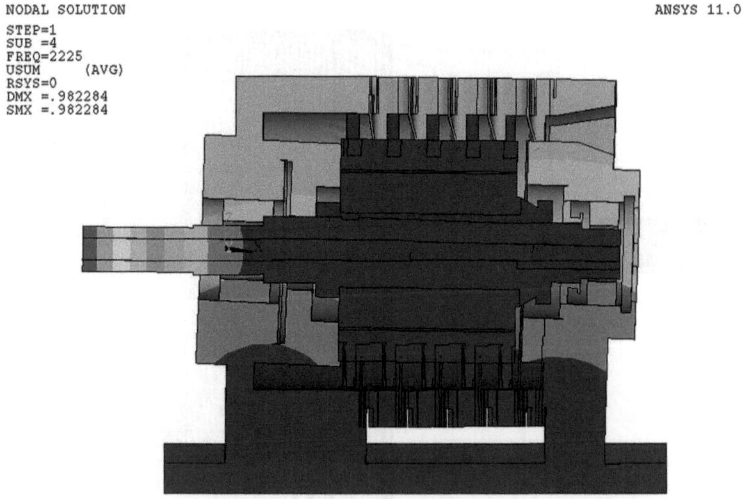

Fig. 4.68 The first form of vibrations of the body at a frequency of 2,225 Hz

They occurred only at very high frequencies and, therefore, they should not occur during normal operation of the machine.

This modal analysis of the axial microturbine confirmed that the tested system was very stiff, thanks to which in the range of operational frequencies, no free vibrations should be induced. This applies to both the rotor and the body, as well as the connection of these two subassemblies.

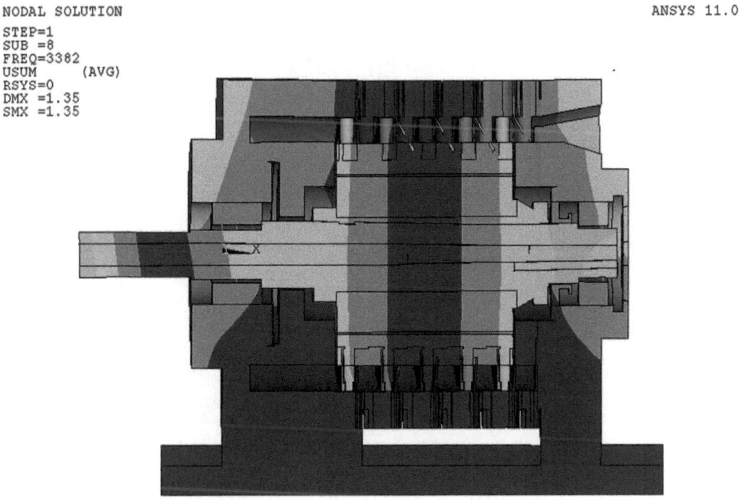

Fig. 4.69 The coupled form of vibrations of the rotor with the body at a frequency of 3,382 Hz (after applying a rigid connection)

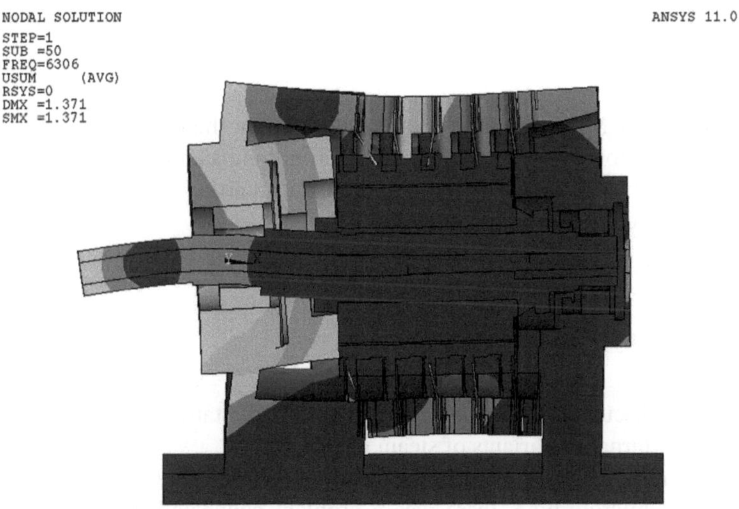

Fig. 4.70 The form of coupled vibrations of the body with the rotor at a frequency of 6,306 Hz (after applying a rigid connection)

4.2.5 Analysis of Kinetostatic Stresses

The strength analysis of the axial microturbine carried out showed that the maximum kinetostatic stresses at a speed of 8,073 rpm were located in the lower part of the friction edge of the rotor blade (Fig. 4.71) [8]. These stresses reached

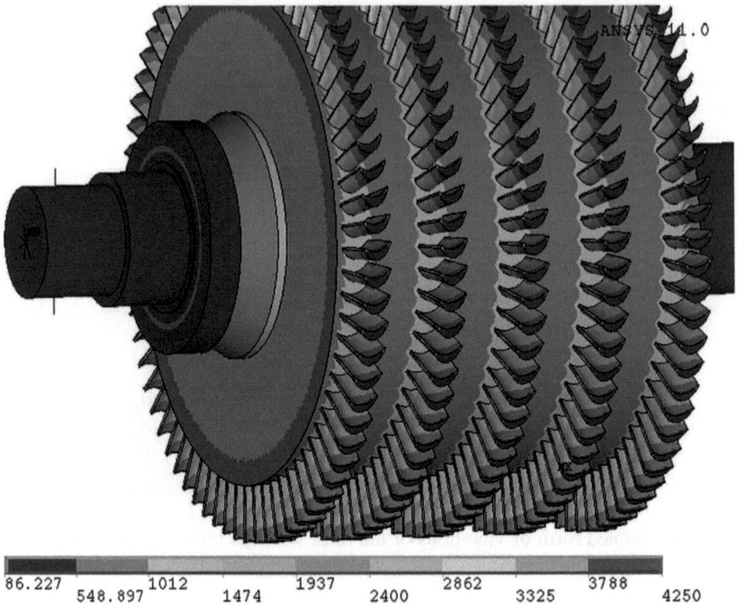

Fig. 4.71 The kinetostatic stresses for the axial microturbine rotor rotating at the nominal speed of 8,073 rpm

approximately 8 MPa and were significantly lower than acceptable stresses of materials used for the manufacture of microturbine elements. It can therefore be concluded that in terms of strength, the axial microturbine can be operated at speeds far exceeding the nominal speed.

4.3 Summary

This chapter discussed the most important results of static, dynamic and strength tests of two alternative variants of steam microturbines.

The dynamic analysis of the rotor-bearings system revealed that in the case of the radial microturbine, there was a problem connected with the occurrence of instability over 12,000 rpm. This problem concerned both hydrodynamic slide bearings and foil bearings lubricated by a low-boiling liquid medium. For the bearings of the rotor of this microturbine, ultimately the use of aerostatic gas bearings was suggested, which then successfully passed the experimental tests. In view of the above, the possible change of the bearing system of the radial microturbine should be preceded by new detailed analyses.

The studies of the axial microturbine rotor-bearings system proved that due to the relatively low rotational speed and beneficial dimensional proportions, it is possible to use both hydrodynamic slide bearings lubricated by a low-boiling medium, as well as rolling bearings. Due to operational conditions of the microturbine, rolling bearings should be made of ceramic materials, which do not require the application of plastic grease, and in the construction of the body, the need to receive heat from the operating bearings should be taken into account.

The computational modal analysis of the radial and axial microturbine carried out showed that both machines worked in subcritical conditions. This is due to the high stiffness of the rotors. In these machines, under normal conditions of use, couplings of vibrations of the rotor with the body also should not occur, nor dynamic problems associated with the vibrations of very stiff blades. Both microturbines also successfully passed the testing strength calculations. The stresses reduced in rotors rotating at the nominal speed were lower than the acceptable stresses. The attempt to accelerate the microturbines (especially radial microturbines) to a speed well above the nominal speed, should be preceded by a new, detailed analysis of the strength.

The numerical models used during the tests were first verified using the results of experimental studies. The methods of conducting such studies according to the modern standards for energy microturbines have been discussed on the example of a number of selected issues in the next chapter.

References

1. Chodkiewicz, R., Klonowicz, P., Łagodziński, J., Więckowski, S.: Design and structural analysis of radial-axial turbine dedicated for CHP power plant operating with low-boiling medium. Int. Study, Institute of Turbomachinery, Lodz University of Technology, Łódź, in Polish (2009)
2. Kozanecki, Z., Więckowski, S., Łagodziński, J., Marczyk, A., Borkowski, R., Kaczmarek, A.: Final assembly, balancing, start-up tests and dynamic tests of the second prototype of turbogenerator with aerostatic bearings using compressed air as the working medium. Int. Study IMP PAN, No. Arch. 59/2011, Gdańsk, in Polish (2011)
3. Kozanecki, Z., Więckowski, S., Łagodziński, J., Marczyk, A., Borkowski, R., Tkacz, E., Kaczmarek, A.: Design and development of technical documentation of the support system for turbogenerator prototype with second-generation foil bearings. Int. Study IMP PAN, No. Arch. 229/2011, Gdańsk, in Polish (2011)
4. Kiciński, J., Żywica, G., Bogulicz, M., Czoska, B., Banaszek, S., Bogucka, K.: Calculation tests of the turbogenerator's rotor variants selected for the further implementation—the analysis of dynamic properties of rotor-bearing system. Int. Study IMP PAN, No. Arch. 71/2010, Gdańsk, in Polish (2010)
5. Kozanecki, Z., Więckowski, S., Łagodziński, J., Marczyk, A., Borkowski, R., Kaczmarek, A.: Dynamic testing of the first prototype of the turbo-generator under maximum load during energy consumption. Int. Study IMP PAN, No. Arch. 671/2010, Gdańsk, in Polish (2010)
6. Kiciński, J., Żywica, G., Bogulicz, M., Czoska, B.: Analysis of the dynamic properties of microturbine rotor supported on foil bearings lubricated with low-boiling medium. Int. Study IMP PAN, No. of arch. 303/2011, Gdańsk, in Polish (2011)
7. Rządkowski, R., Drewczyński, M., Kiciński, J., Żywica, G.: Calculation tests of the turbogenerator's rotor variants selected for the further implementation—modal analysis. Int. Study IMP PAN, No. Arch. 113/2010, Gdańsk, in Polish (2010)
8. Rządkowski, R., Drewczyński, M., Kiciński, J., Żywica G.: Analysis of free vibrations and kinetostatic stresses of the rotor and radial microturbine body for the full range of rotational speeds. Int. Study IMP PAN, No. Arch. 386/2011, Gdańsk, in Polish (2011)

9. Łączkowski, R.: Vibrations of Components of Gas and Steam Turbines. WN-T, Warsaw, in Polish (1974)
10. Kiciński, J., Żywica, G., Bogulicz, M.: Analysis of the impact of selected design and operating parameters on the operating conditions of the rotor-bearing system in steam microturbine with ORC. Int. Study IMP PAN, No. Arch. 133/2012, Gdańsk, in Polish (2012)
11. Kiciński, J., Żywica, G.: Development of integration concept of basic components of the small CHP power plant with ORC. Schedule of construction and integration works. Int. Study IMP PAN, No. Arch. 588/2011, Gdańsk, in Polish (2011)
12. Kiciński J.: Dynamics of rotors and slide bearings. IMP PAN Publication, Gdańsk, in Polish (2005)
13. Żywica, G.: The analysis of the defects of the supporting structure in relation to the dynamic state of the turbomachine. Ph.D. thesis, IMP PAN, Gdańsk, in Polish (2011)
14. Piwowarski, M., Kosowski, K.: Determination of the main design parameters and constructional type of the turbine. Int. Study IMP PAN, No. Arch. 215/2009, Gdańsk, in Polish (2009)
15. Kosowski, K., Piwowarski, M., Stępień, R., Włodarski, W.: The design of flow part of the turbine. Int. Study IMP PAN, No. Arch. 838/09, Gdańsk, in Polish (2009)
16. Kosowski, K., Piwowarski, M., Stępień, R., Włodarski, W.: Strength calculation tests of selected microturbine elements. Int. Study IMP PAN, No. Arch. 839/09, Gdańsk, in Polish (2009)
17. Institute of Innovative Technology: Design and construction analysis of the axial, multistage turbine with partial power supply being a specially dedicated solution for small CHP power plant operating with low-boiling medium. Study commissioned by IMP PAN, Gdańsk, in Polish (2009)
18. Kiciński J., Żywica G., Bagiński P.: Analysis of the dynamic properties of the rotor-bearing system in axial microturbine with rolling bearings. Int. Study IMP PAN, No. Arch. 450/2010, Gdańsk, in Polish (2010)
19. Krzemiński-Freda, H.: Rolling Bearings. PWN, Warsaw, in Polish (1989)
20. Rządkowski, R., Drewczyński, M., Kiciński, J., Żywica, G.: Analysis of the dynamic properties of turbogenerator's rotor in the small steam power plant taking into account the initial geometry of the frame. Int. Study IMP PAN, No. Arch. 296/2010, Gdańsk, in Polish (2010)

Chapter 5
Measurements of Vibrations of Microturbine Elements

Abstract This chapter focuses on experimental studies of various elements of steam microturbines. At the beginning of the chapter, experimental modal analysis of the microturbine's rotor was discussed in detail and the results were compared with the results of computational modal analysis. The results are presented in the form of stabilization diagrams, free vibration modes and frequencies. Other studies concerned the evaluation of the dynamic properties of the rotor with ceramic rolling bearings. The results were compared with the characteristics of classic (steel) rolling bearings. The following sections discuss the research on steam microturbines powered by steam of low-boiling medium. During the experimental study of the radial microturbine the symptoms had been observed indicating the existence of defects, which was also described in this chapter. A review of the collected data and thorough observation allowed for early detection of defects as well as finding the causes of their occurrence. Next, the study of the dynamic properties of innovative construction of the supersonic microturbine and an assessment of its dynamic state were discussed. The last part of this chapter presents the study of the dynamic properties of the microturbine's supporting structure, in this case the aluminium frame. The test results presented in this chapter are a very valuable material that can be used by other researchers, e.g. to compare dynamic properties of similar design solutions, or to verify numerical models.

This chapter discusses issues related to the experimental studies of steam microturbines and their subassemblies. Due to the very large number of experimental test results obtained during the realization of the project, only selected effects of the measurements done will be presented here, which are representative of a certain class of relevant issues from the point of view of the dynamics of energy microturbines. The authors of the book refer readers who are interested in the details of these issues to the cited works and available literature.

J. Kiciński and G. Żywica, *Steam Microturbines in Distributed Cogeneration*,
DOI 10.1007/978-3-319-12018-8_5

5.1 Experimental Modal Analysis of the Rotor

This part of the chapter discusses the course of studies and presents the results of an experimental modal analysis of the radial microturbine rotor [1], which was discussed in detail in Sect. 4.1. The studies were carried out in order to obtain experimental data allowing to verify the developed numerical models of the rotor, used for dynamic and strength analysis.

The rotor in the assembled microturbine consisted of several elements (Fig. 5.1). The fundamental part was the shaft, on which two discs were embedded (rotor disc and longitudinal bearing disc) secured using threaded connections. Additionally, in the central part of the shaft the generator sleeve was mounted. The designed shaft had two polished journals adapted to working with two transverse aerostatic bearings.

For the needs of carrying out experimental modal tests, the microturbine rotor was simplified in accordance with Fig. 5.2. The tested rotor, while maintaining the primary dimensions had a simplified geometry. It also did not include the disc of the longitudinal bearing, which in one of the variants developed in the course of the implementation of the project was removed and replaced by two longitudinal bearings integrated with transverse bearings. The generator sleeve was also not included. Due to the very high frequencies of vibrations of the rotor blades (far outside the range of the operation of microturbines [2]), these parts were not made in the model of the rotor used for modal testing. The rotor shield without blades was permanently fixed on the shaft, and the whole mass of the rotor was 1,482 g.

For auxiliary purposes, before performing an experimental modal analysis, a simulation modal analysis of the system was carried out. For this variant of the radial microturbine, the nominal rotational speed was 23,800 rpm. This speed corresponds to the rotational frequency equal to 396.7 Hz. The first bending mode for the rotor vibrations indicated by simulation using the ABAQUS program, including all its elements (according to Fig. 5.1) amounted to 1,405 Hz (Fig. 5.3), and so

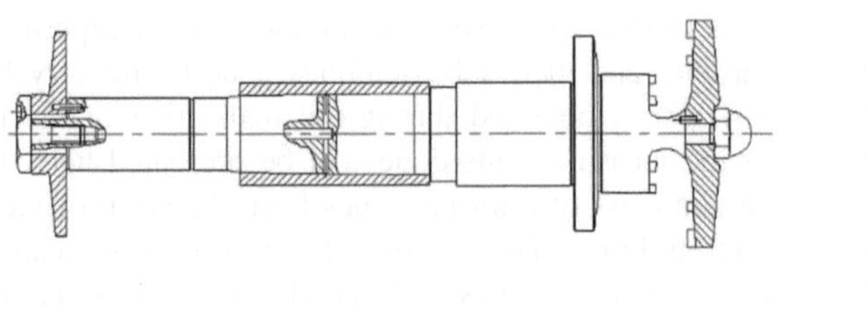

Fig. 5.1 Drawing of the radial microturbine rotor

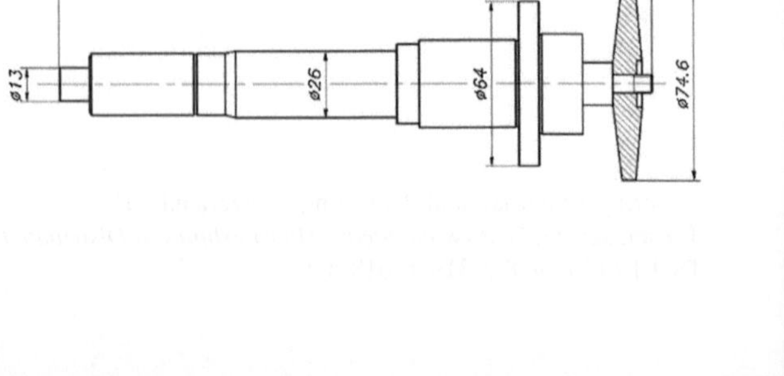

Fig. 5.2 Drawing of the simplified radial microturbine rotor

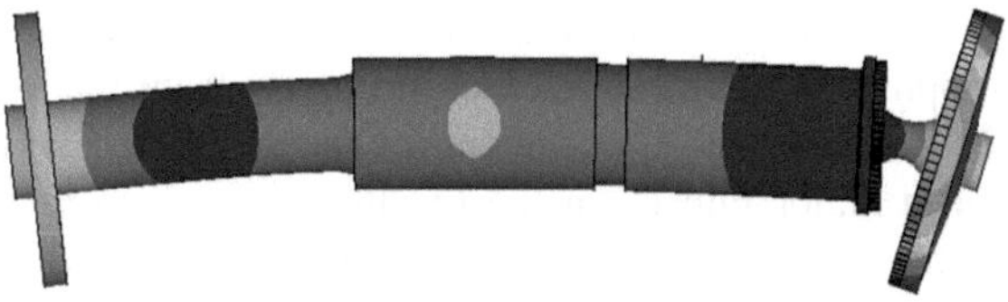

Fig. 5.3 The first bending mode of the microturbine rotor determined by simulation (1,405 Hz)

it was more than three times greater than the frequency corresponding to the nominal speed. Subsequent bending modes of the rotor, determined through computer simulation, occurred at the frequencies 2,081 and 4,392 Hz.

Before the experimental tests were begun, calculations were also performed for the simplified rotor of the modelled radial microturbine, which has been shown in Fig. 5.2. The simulation modal analysis demonstrated that the first bending mode after implementing simplifications occurred at a frequency of 1,976 Hz (Fig. 5.4). The second and third bending mode of this rotor occurred at frequencies 3,377 and 6,414 Hz. These modes have been presented in Figs. 5.5 and 5.6.

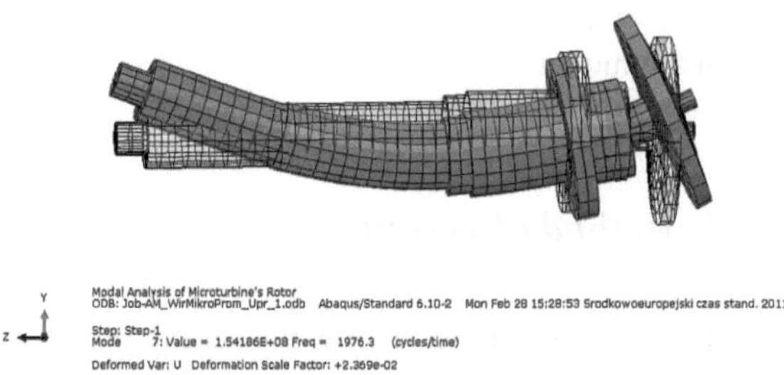

Fig. 5.4 The first bending mode of the simplified microturbine rotor determined by simulation

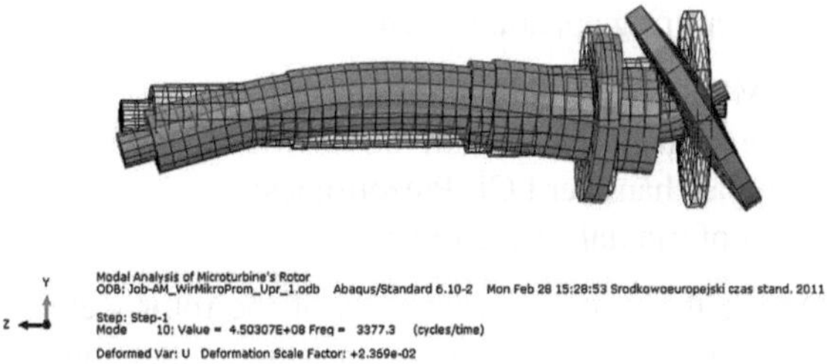

Fig. 5.5 The second bending mode of the simplified microturbine rotor determined by simulation

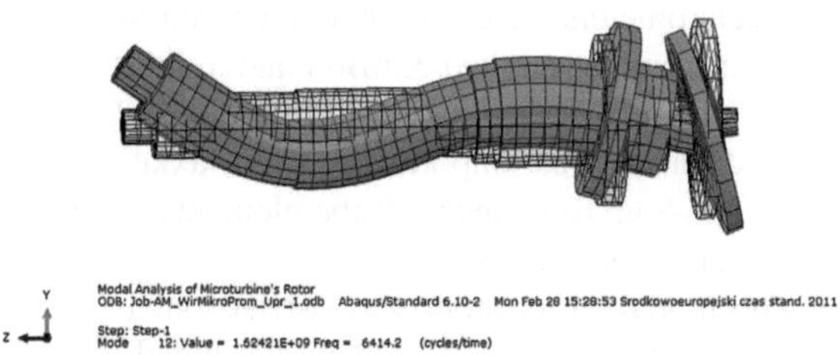

Fig. 5.6 The third bending mode of the simplified microturbine rotor determined by simulation

The differences in frequencies of the first bending mode, which occurred after simplifying the rotor, resulted primarily from the failure to include the longitudinal bearing, placed at the end of the shaft. The way the rotor was mounted also had a major impact on the results obtained. In the case of previous studies, the stiffness of bearings was taken into account, and the results presented in Figs. 5.4, 5.5 and 5.6 were obtained for the freed model.

According to the results of the simulation modal analysis, the tested rotor worked in the range of subcritical rotational speeds. Due to the high stiffness of the rotor, the first bending mode occurred outside the range of operational frequencies. This meant that its dynamic properties will mostly be affected by the characteristics of the bearings, and not the modal properties of the rotating system. Therefore, bending vibrations will not be induced as a result of operational loads, such as residual unbalance or forces occurring in the generator. The dynamic condition of the operating system will also not be affected by the vibrations of rotor blades, the free vibration modes of which could be observed at frequencies exceeding 50 kHz.

For these reasons, and in view of the typical limitations of modal measurements, during experimental studies, the focus was on the identification of the first bending mode of the microturbine rotor [3]. For it was possible to measure it when using available testing apparatus, and thus, it was possible to carry out a verification of the model.

5.1.1 Method of Testing

During these modal tests, the measuring apparatus that was available in the Laboratory of Vibrodiagnostics was used, which belongs to the Dynamics and Diagnostics of Turbines Department (IMP PAN). The most important elements of the measuring apparatus were:

- LMS SCADAS Mobile Analyzer;
- Personal computer with the software LMS Test.Lab;
- Impact hammer PCB Piezotronics;
- Set of triaxial accelerometers.

During the tests, the vibrations of the rotor were driven by a hammer for modal tests with a mass of 0.16 kg, in which a metal end was used allowing to drive the system in the frequency range up to 10 kHz. To measure the vibrations, two types of accelerometers were used manufactured by PCB Piezotronics: integrated triaxial accelerometer marked 356B08 and a uniaxial accelerometer marked 080A140. Due to the relatively low weight of the tested system, the mass of accelerometers applied was of great importance. The triaxial accelerometer weighed 38 g, and the uniaxial—6 g. To connect all the elements of the measuring system, high-quality electrical wires were used.

The modal experiment was carried out in such a way that the vibrations of the system were driven by a hammer, which was used to strike near the uniaxial accelerometer. The triaxial measurement of the system's response to the drive caused, took place in 20 measuring points on the outer surface of the rotor, in five parallel planes, which were perpendicular to the axis of the rotor. The location of measuring points was proposed on the basis of preliminary findings of the simulation modal analysis, taking into account the distributions of nodes and arrows of vibrations. The distribution of adopted measuring points has been shown in Fig. 5.7.

In order to minimize the impact of support method, the rotor was suspended using an elastic tie (Fig. 5.8). This suspension was to model the free-free type border conditions. During the preparations for tests, mounting accelerometers to the rotor turned out to be problematic. In fact, it was difficult to obtain a durable and

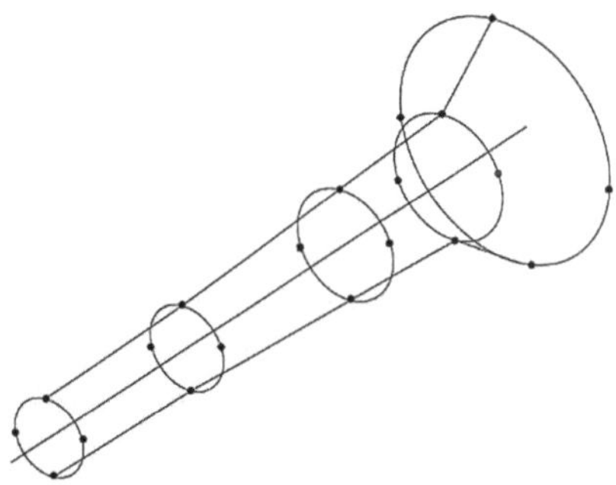

Fig. 5.7 The distribution of measuring points on the tested rotor

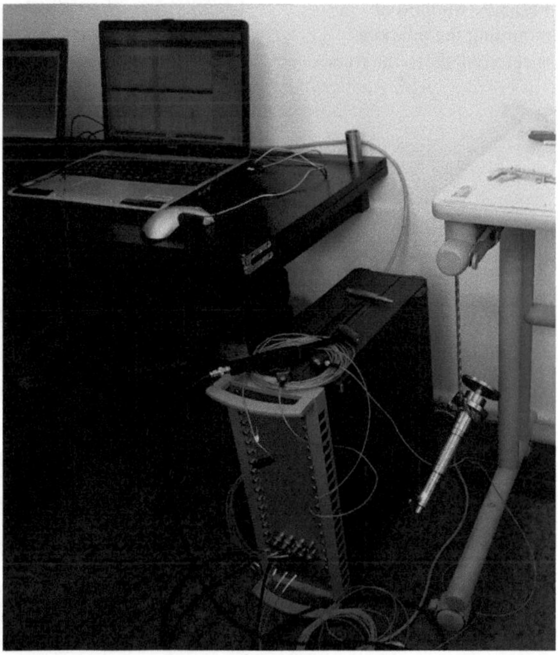

Fig. 5.8 The view of the position at which measurements took place

easily removable connection between the flat surface of the sensor and surface of the rotor with a small curvature radius. Mounting the accelerometer exclusively on the magnetic base caused it to swing on the cylindrical surface of the rotor. Therefore, in the case of a uniaxial accelerometer (mounted close to the location of inducing vibrations), the sensor was glued to the tested element using a two-ingredient epoxy resin. In order to facilitate disassembly, the accelerometer and shaft were first covered with aluminium foil, and the glue did not have direct contact with the connected elements (Fig. 5.9). In the rest of the measuring points, the triaxial accelerometer was mounted using a magnetic base, as during the tests it was necessary to move it to subsequent measuring points.

Due to the relatively large mass of the triaxial accelerometer in relation to the mass of the rotor (the mass of the sensor was approx. 2.5 % of the mass of the tested object), the idea of application of a larger number of sensors was abandoned. Using them would contribute to reducing the time of measurements, but at the expense of accuracy of the results. The method of fixing the triaxial piezoelectric accelerometer has been shown in Fig. 5.10. This diagram also shows other measuring points, which have been marked by a red marker.

The measurement of the vibrations of the tested rotor consisted of a total of 19 measurement series, in which measurements were made and the registration of the vibrating system's responses in subsequent measuring points. In each measurement series 10 impact hammer strokes were made near the uniaxial accelerometer (Fig. 5.11). All the imprecise and too strong strokes (causing signal distortion on any measuring track) were automatically rejected and needed to be repeated.

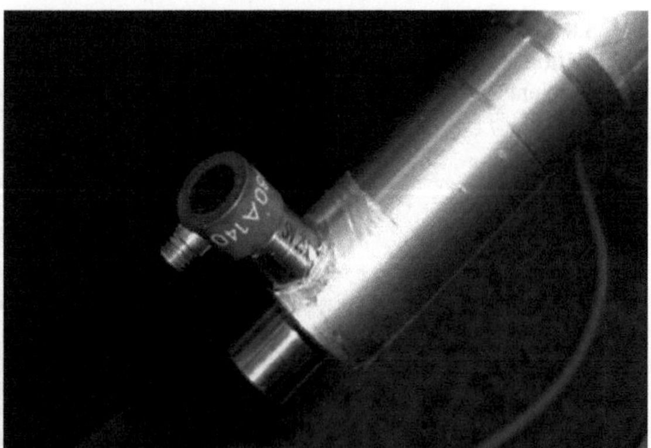

Fig. 5.9 Method of mounting the uniaxial accelerometer to the rotor

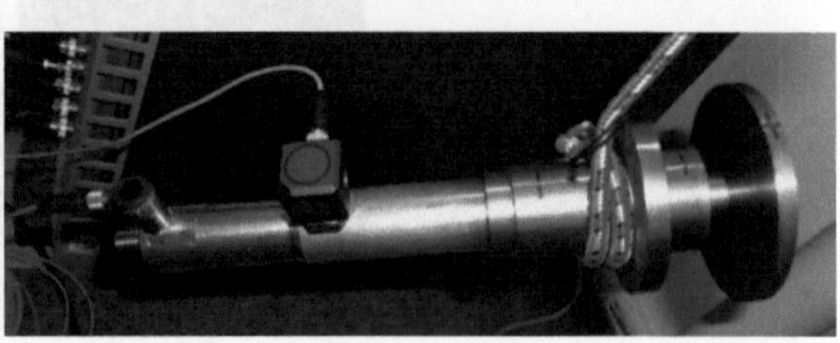

Fig. 5.10 The method of mounting the triaxial piezoelectric accelerometer in the selected measuring point

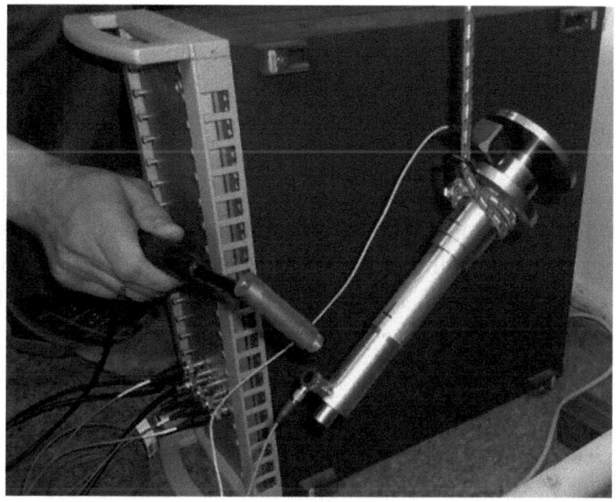

Fig. 5.11 Driving vibrations of the rotor using an impact hammer

5.1.2 Test Results

After conducting the measurements, on the basis of the recorded waveforms, a stabilization diagram was created, which enabled the selection of the poles in which modes of free vibration could occur. Then, for these frequencies, diagrams were drawn up and animations created which demonstrated the vibrations of the rotor.

The basis for estimating modal parameters was the stabilization diagram shown in Fig. 5.12. The range of analyzed frequencies was limited to the values between 1,140 and 8,600 Hz. The polynomial degree used to plot the curve visible in the stabilization diagram was 32. In accordance with the presented diagram, there are a few maximums visible on the obtained curve, however, only the first maximum,

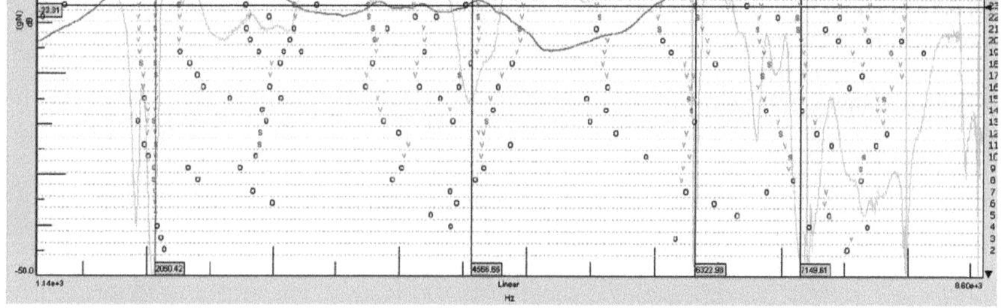

Fig. 5.12 The stabilization diagram with indicated poles of stability of results

occurring at 2,060 Hz clearly shows the presence of free vibration modes of the system. Local maximums occurring at higher frequencies have been treated as potential frequencies, at which free vibration modes could occur. The presented diagram indicated only stable poles due to the frequency, modal vector and damping coefficient (marked by the letter s). A total of 4 frequencies were used for further analysis: 2,060, 4,567, 6,323 and 7,150 Hz, for which a detailed assessment was performed on the basis of visualizations and animations of the system's vibrations.

The following drawings (Figs. 5.13, 5.14, 5.15 and 5.16) show a graphic interpretation of the tested rotor's vibrations for frequencies indicated in the

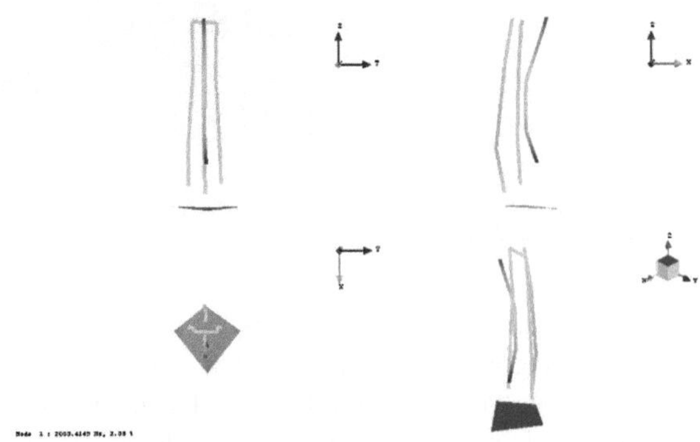

Fig. 5.13 Graphic representation of the rotor's vibrations at a frequency of 2,060 Hz (damping 2.08 %)

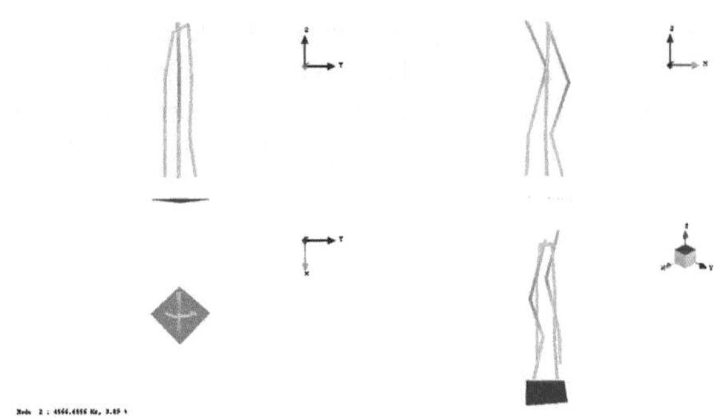

Fig. 5.14 Graphic representation of the rotor's vibrations at a frequency of 4,567 Hz (damping 3.89 %)

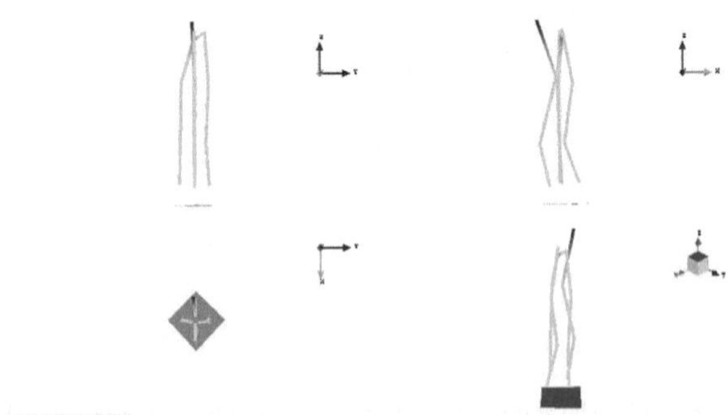

Fig. 5.15 Graphic representation of the rotor's vibrations at a frequency of 6,323 Hz (damping 2.91 %)

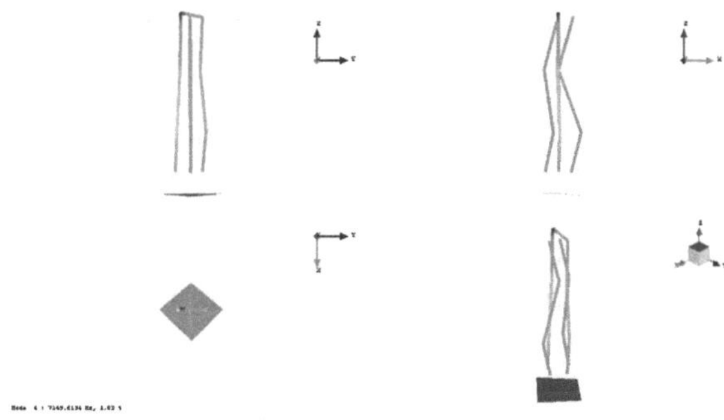

Fig. 5.16 Graphic representation of the rotor's vibrations at a frequency of 7,150 Hz (damping 1.83 %)

stabilization diagram. Because the drive of the system took place in the XZ plane, in the following drawings, particular focus should be given to the results obtained in this plane.

The vibration mode presented in Fig. 5.13 is evidently the first bending mode of the tested rotor. It occurred at a frequency of 2,060 Hz with damping equal to 2.08 %. The same mode of free vibration was obtained by simulation at a frequency equal to 1,976 Hz. The difference of frequencies between the result of the experiment and simulation was therefore only 4.08 %. It should be emphasized that the numerical model applied had not been previously subjected to experimental verification or tuning, therefore, the initially obtained compliance of results was very good.

The interpretation of the results obtained at higher frequencies, however, is not as clear. Due to the high convergence of frequencies and the similar shape of the deflected rotor line, the vibrations presented in Fig. 5.15 could be interpreted as a third bending mode. At other frequencies indicated in the stabilization diagram there were vibrations, which were difficult to accurately identify. Also in the results of the simulation modal analysis, in the area of indicated frequencies, no modes of free vibration were observed. The problems encountered with the interpretation of the results are quite common in the case of experimental modal analysis. In these tests, the relatively large mass of the applied vibration sensors, as well as the method of mounting the triaxial accelerometer definitely had a negative effect on the quality of the results obtained.

The results of the experimental modal analysis confirmed the fact that due to the high stiffness of the rotor, the first bending mode occurred well above the provided rotational frequencies of the system. Because the tested rotor is subcritical, its stable operation will be mainly affected by the stiffness-damping properties of the bearings.

Due to the fact that during the conducted modal studies it was possible to clearly identify only the first bending mode of the rotor's vibration, it would be desirable to carry out further studies. Another way to verify the applied numerical models could be the use of the results of the measurement of rotor vibrations in the operating microturbine. Additionally, such tests allow the verification of the models of bearings used. Further parts of this chapter deal with these types of tests.

5.2 Vibrations of the Rotor with Ceramic Rolling Bearings

Due to the difficult operational conditions and airtight construction, steam microturbines require the use of unconventional bearing systems. One of the modern, rarely used methods of bearing, are ceramic rolling bearings, which are the subject of the studies described in this part of the book. The most important objective of these works was to carry out experimental tests of a rotor with ceramic rolling bearings, as well as an assessment of using these bearings in one of the developed structural solutions of the prototype steam microturbine [4]. With regard to microturbines, the basic advantages of rolling bearings made from ceramic materials are:

- the ability to work at high rotational speeds;
- resistance to high temperatures;
- no need to use conventional lubricant mediums.

This makes it possible to put forward the thesis that ceramic rolling bearings can be a very good alternative to other, advanced bearing systems of energy microturbine rotors.

5.2.1 *Method of Testing*

The configuration and basic dimensions of the position in which the tests were performed have been shown in Fig. 5.17. For the purposes of these tests, a shaft

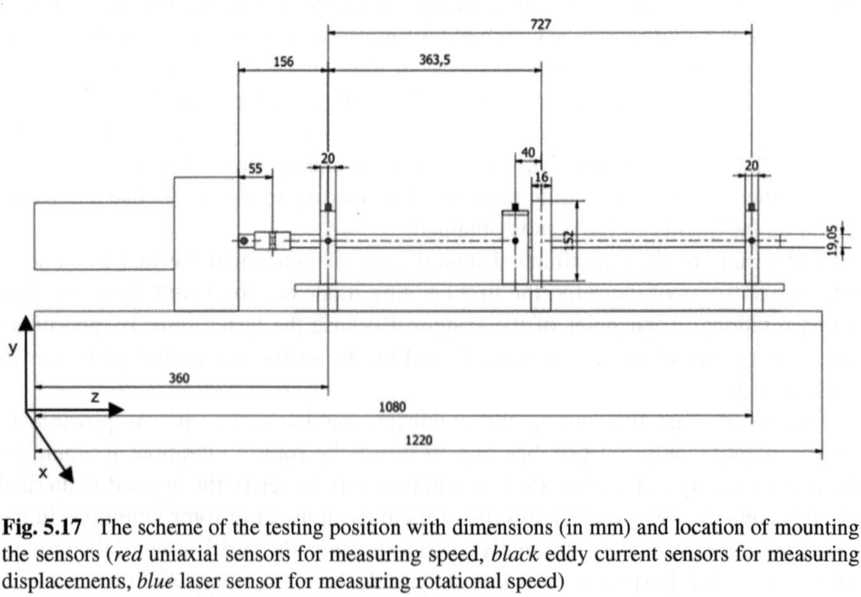

Fig. 5.17 The scheme of the testing position with dimensions (in mm) and location of mounting the sensors (*red* uniaxial sensors for measuring speed, *black* eddy current sensors for measuring displacements, *blue* laser sensor for measuring rotational speed)

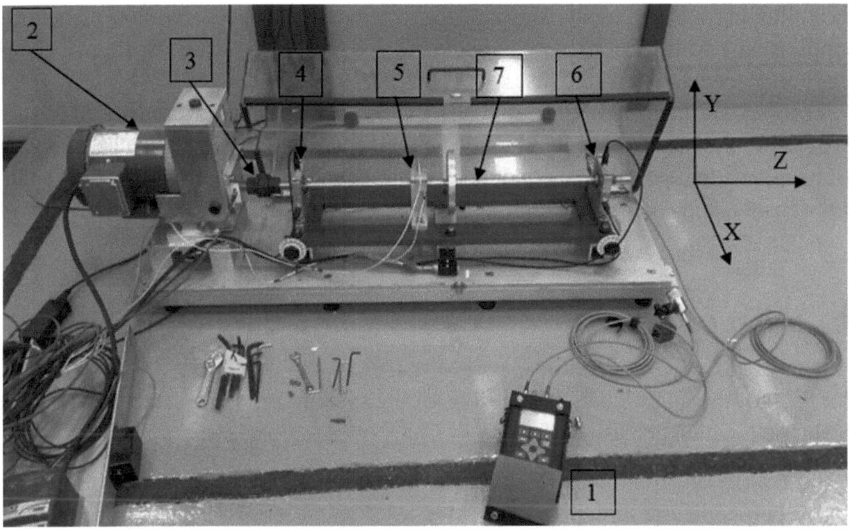

Fig. 5.18 Photograph of the test rig (*1* mobile vibration analyzer, *2* engine with gearbox, *3* elastic clutch, *4* first bearing support, *5* support of eddy current sensors, *6* second bearing support, *7* tested rotor)

and rotor disc were made with a total weight of 3.5 kg. The power drive of the position was connected with the shaft using an elastic clutch (Fig. 5.18). Uniaxial accelerometers were fixed in the bearing supports in the directions of the axes X and Y. The rotational speed of the rotor was measured using a laser sensor. The displacements of the shaft at the rotor disc were measured in two perpendicular directions, using eddy current sensors with own signal conditioning.

The whole apparatus was connected to a SCADAS Mobile analyzer, which processed and recorded signals from all the sensors and connected with a computer workstation. All of the measurement parameters were configured using the software Test.Lab with the package Signature Testing [5]. The software used enabled also to validate and visualize the results obtained.

In addition to the entirely ceramic rolling bearings, for comparison purposes, rolling bearings made from steel and hybrids (ceramic-steel) were also examined. All the bearings had the same nominal diameter of 20 mm. After every change of bearings, the rotor was balanced again using the mobile analyzer of vibrations Diamond 401 with the function of balancing shafts in own bearings. Balancing consisted in setting an appropriate corrective mass, as indicated by the analyzer of vibrations, which lead to reducing the level of vibrations. It was considered that a rotor was sufficiently well-balanced if the effective value of vibration speed V_{RMS} reached approx. 1 mm/s.

When preparing the measurements, it turned out that the correct setting of eddy current sensors was a big problem, as their measurement range was limited to 0.5–2.5 mm. Therefore, they were set up in such a way, so that their position at a resonant speed did not allow damage (abrasion) to occur. During the performed tests, the following measurements were made: rotational speed, acceleration of bearing supports in two directions perpendicular to the axis of rotation of the rotor, and shaft displacements near the disc.

For each of the tested types of bearings, the measurements were carried out for two cases: a balanced and deliberately unbalanced rotor. This sequence of measurements made it possible to conduct tests at unchanged configuration of eddy current sensors, and enabled the comparison and evaluation of the results obtained.

5.2.2 Test Results

For each of the analyzed configurations of the system, the same set of results were developed, which consisted of:

- the course of amplitudes of vibration acceleration in the bearing supports (in two perpendicular directions);
- cascading graphs of the course of vibration acceleration and so-called maps of colours registered in the bearing supports (also in two perpendicular directions);
- shaft displacements near the disc;
- trajectories of vibrations of bearing supports and the shaft for a few selected rotational speeds.

Figure 5.19 below provides the courses of amplitudes of vibration acceleration for the rotor with ceramic bearings in a balanced state, recorded on two supports. The presented graph makes it possible to distinguish three speeds, for which an increase in the vibration acceleration amplitude occurred: 3,550, 4,916, 8,900 rpm. For the first support in X-direction, a gradual increase in vibration

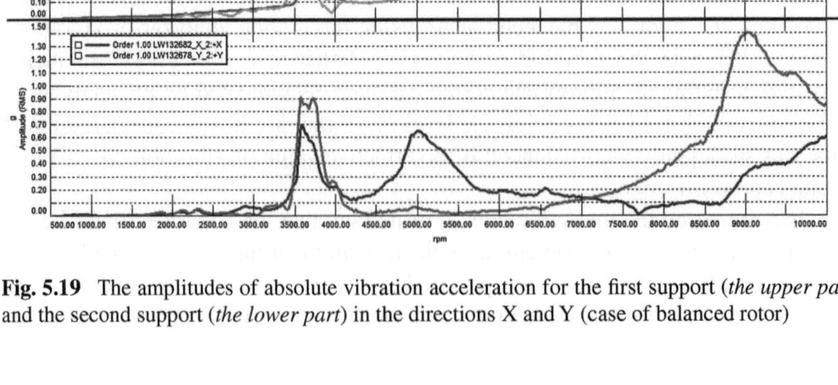

Fig. 5.19 The amplitudes of absolute vibration acceleration for the first support (*the upper part*) and the second support (*the lower part*) in the directions X and Y (case of balanced rotor)

acceleration was observed at high speeds of the rotor, until it reached the value of 1.2 g at maximum speed. In the first support in Y-direction, the maximum acceleration value was 0.75 g at 8,900 rpm. In the second support in X-direction, the vibration acceleration did not exceed 0.7 g, and in Y-direction it reached the level of 1.41 g (also at 8,900 rpm).

The graphs presented in Figs. 5.20, 5.21, 5.22 and 5.23 show the frequency distributions of vibration accelerations recorded in the bearing supports at different rotational speeds. Vibrations with a frequency compliant with rotational

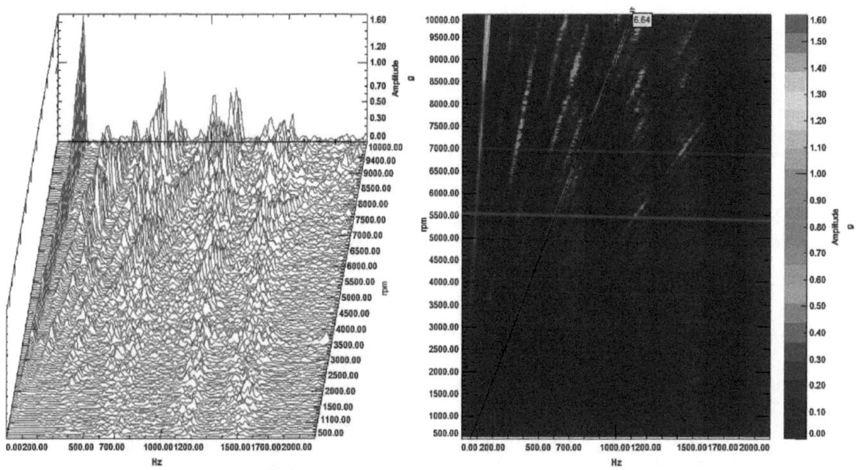

Fig. 5.20 Cascade graph (*left side*) and map of *colours* (*right side*) for a uniaxial acceleration sensor from the first support placed in Y-direction (case of the balanced rotor)

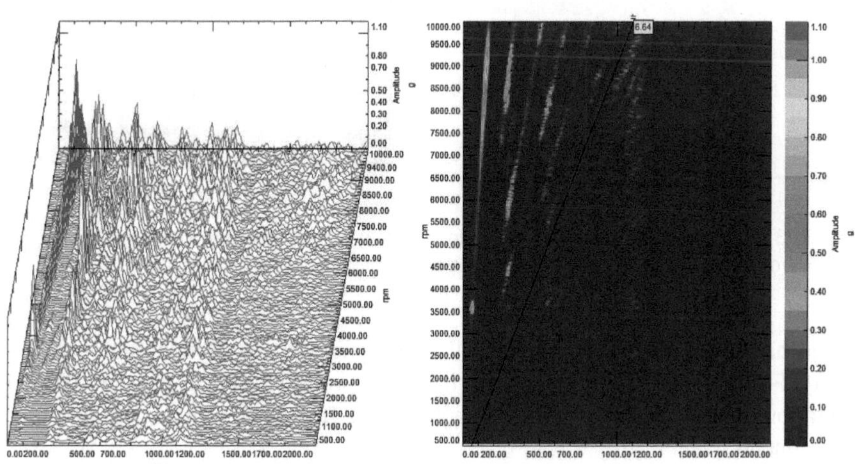

Fig. 5.21 Cascade graph (*left side*) and map of *colours* (*right side*) for a uniaxial acceleration sensor from the first support placed in X-direction (case of the balanced rotor)

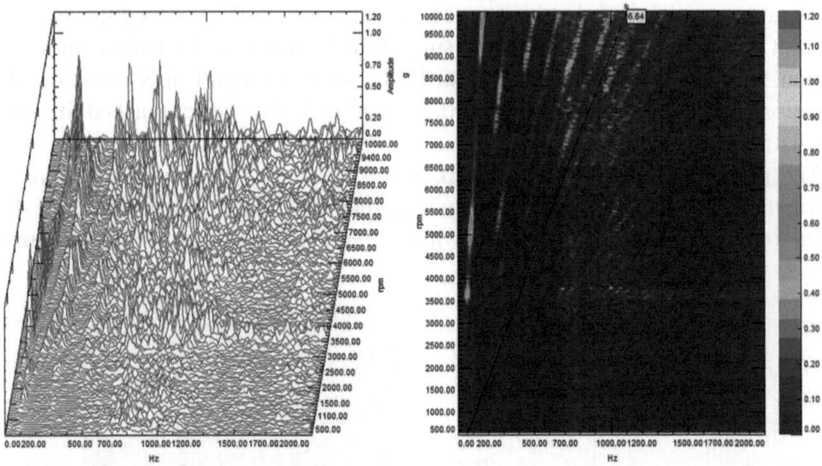

Fig. 5.22 Cascade graph (*left side*) and map of *colours* (*right side*) for a uniaxial acceleration sensor from the second support placed in X-direction (case of the balanced rotor)

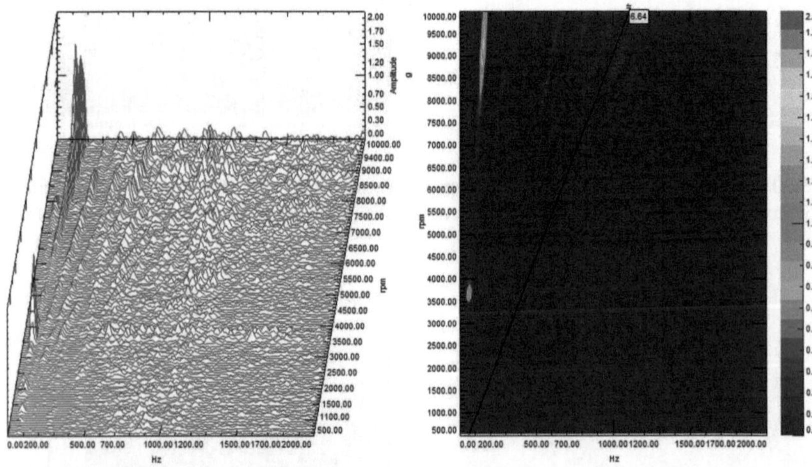

Fig. 5.23 Cascade graph (*left side*) and map of *colours* (*right side*) for a uniaxial acceleration sensor from the second support placed in Y-direction (case of the balanced rotor)

speed had a dominating impact on all four characteristics (synchronous component). Furthermore, in both supports in the recorded signal, an elevated level of vibrations could be observed corresponding to frequencies of higher harmonic components.

The next Fig. 5.24 provides the courses of the peak-to-peak amplitude of shaft displacement near the disc (RMS value) depending on the rotational speed. The signal was measured using eddy current sensors with a conditioning system. The

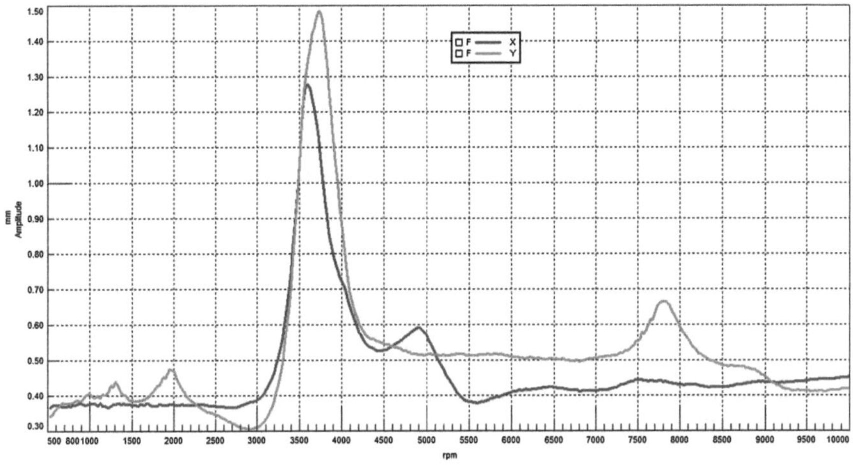

Fig. 5.24 Amplitudes of rotor displacements depending on the speed, measured at the disc in the direction of X and Y (case of balanced rotor)

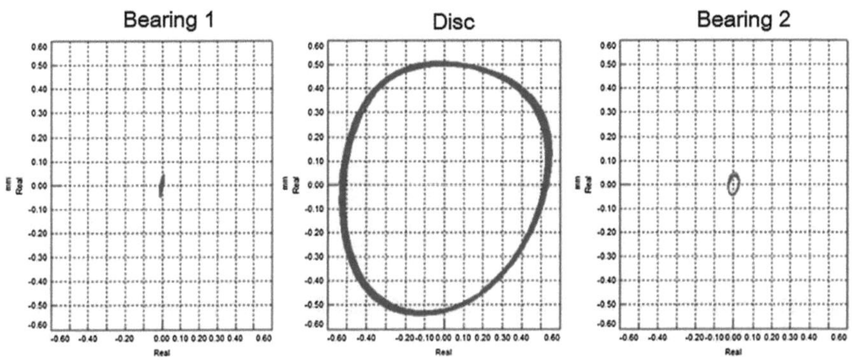

Fig. 5.25 The trajectories of absolute vibrations of a balanced rotor with ceramic bearings at a speed of 3,500 rpm

presented characteristics show that at a low speed the rotor worked steadily. Below the resonant speed there were small values of vibration amplitudes observed. In the presented characteristics certain differences can also be seen between the maximum displacements in the direction of X and Y. In X-direction the maximum displacement amounted to 1.3 mm for a speed of 3,580 rpm, while in Y-direction the greatest displacement amounted to as much as 1.5 mm at 3,755 rpm.

The following Figs. 5.25, 5.26 and 5.27 show the trajectories of the rotor's vibrations measured near the disc at speeds of 3,500, 4,000 and 8,000 rpm. At a speed of 3,500 rpm the system was before resonance, and at 4,000 rpm after crossing resonance. The trajectories at quite a high speed of 8,000 rpm have additionally been presented. On the basis of measurements performed, it can be concluded

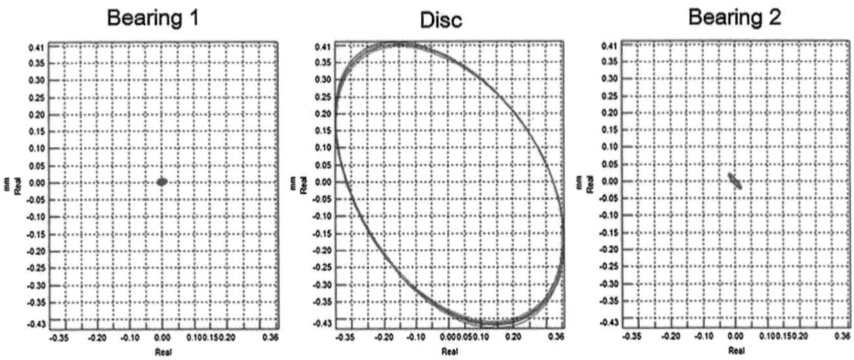

Fig. 5.26 The trajectories of absolute vibrations of a balanced rotor with ceramic bearings at a speed of 4,000 rpm

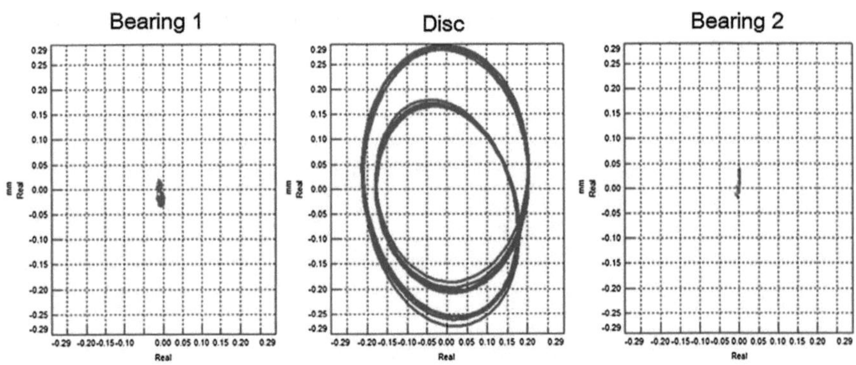

Fig. 5.27 The trajectories of absolute vibrations of a balanced rotor with ceramic bearings at a speed of 8,000 rpm

that the tested rotor with ceramic bearings in most speed ranges worked steadily. Only at higher speeds (above 7,000 rpm) a split in the trajectories of vibrations was observed, and the harmonic analysis of the signal showed the appearance of subsynchronous components. At these speeds irregular trajectories of vibrations were observed, mainly at bearing supports.

In the remainder of this subsection, the results of the measurements obtained for the same system are presented, after deliberate introduction of unbalance (on the outer surface of the disc an additional mass of 0.85 g was mounted). The use of extra mass in many cases enables to obtain clearer characteristics and evaluation of the system in conditions in which a known dynamic load operates.

Figure 5.28, showing the amplitudes of vibration acceleration, as many as five areas can be distinguished at which an increase occurred in the level of recorded vibration signal: 3,480, 3,888, 4,870, 6,576, 8,910 rpm. For the first support in X-direction the recorded curve of vibration acceleration had an almost constantly

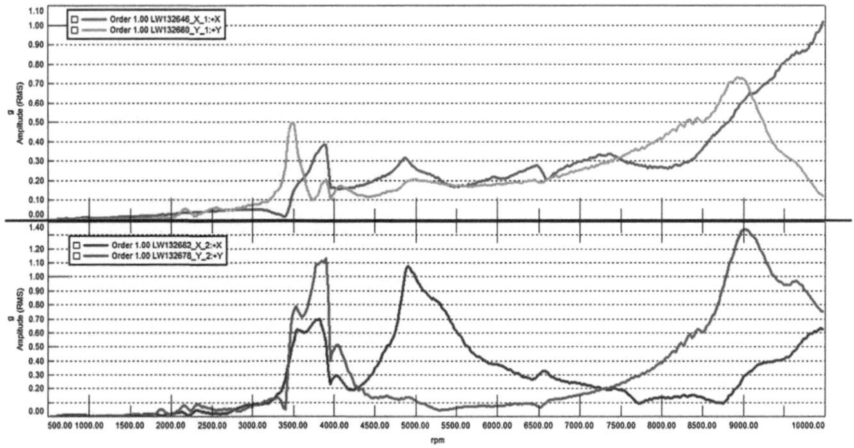

Fig. 5.28 The amplitudes of absolute vibration acceleration for the first support (*the upper part*) and the second support (*the lower part*) in the directions X and Y (case of rotor with added mass)

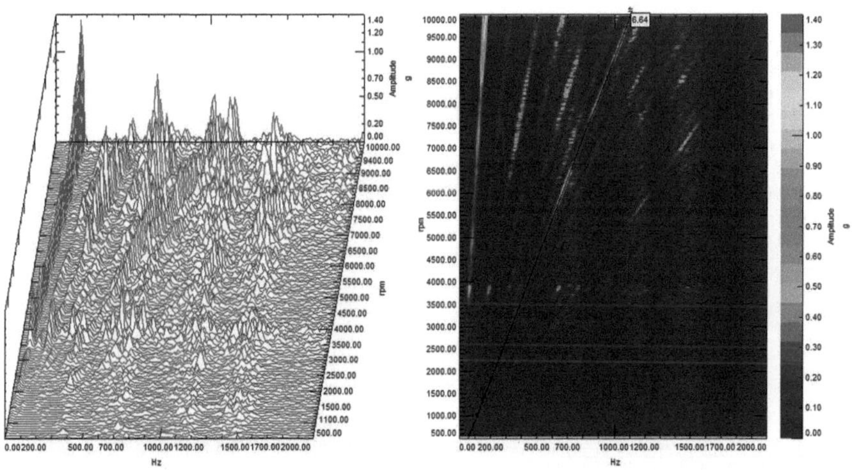

Fig. 5.29 Cascade graph (*left side*) and map of *colours* (*right side*) for a uniaxial acceleration sensor from the first support placed in X-direction (case of the unbalanced rotor)

growing character and reached a maximum of 1.02 g. In Y-direction the largest amplitude amounted to 0.76 g at 8,910 rpm. For the second support in X-direction the maximum value recorded was 1.08 g, while in Y-direction as much as 1.36 g. Compared to the results obtained for the balanced rotor (Fig. 5.19) in Fig. 5.28 additional areas with an elevated level of vibration acceleration have become visible.

The following drawings show the cascade type graphs and maps of colours for two bearing supports (Figs. 5.29, 5.30, 5.31 and 5.32). For the first support in X-direction (Fig. 5.29) a clear distinction of a few vibration components can

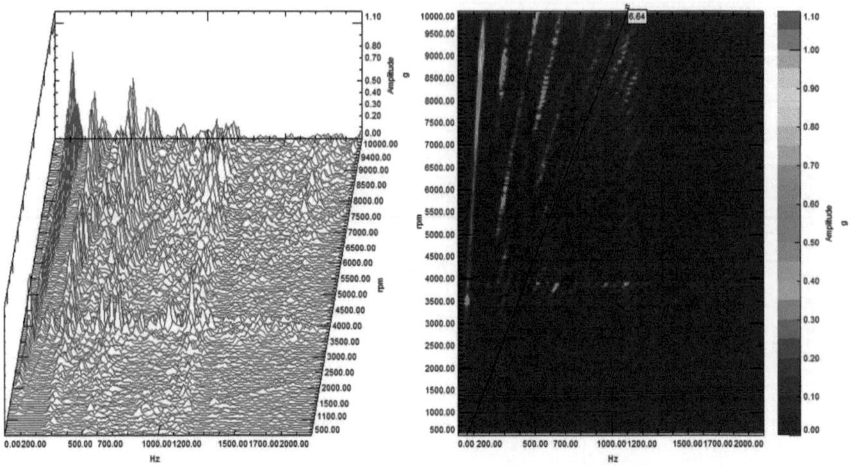

Fig. 5.30 Cascade graph (*left side*) and map of *colours* (*right side*) for a uniaxial acceleration sensor from the first support placed in Y-direction (case of the unbalanced rotor)

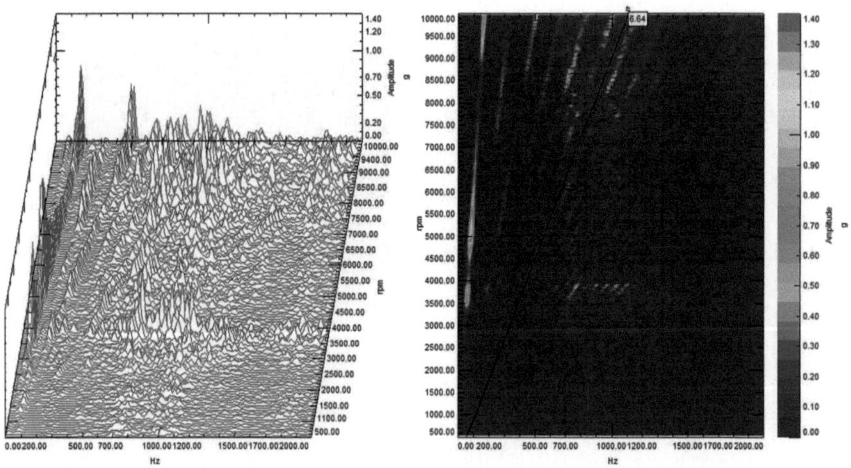

Fig. 5.31 Cascade graph (*left side*) and map of *colours* (*right side*) for a uniaxial acceleration sensor from the second support placed in X-direction (case of the unbalanced rotor)

be observed. In most of the range of rotations the first component was dominant, but in terms of higher speeds (6,750–8,580 rpm) its exceeding by a third harmonic component could be observed. In Y-direction (Fig. 5.30) a similar situation occurred, but the dominance of the third harmonic component was observed in the speed range 4,976–6,422 rpm. In light of the fact that the graphs presented below show the presence of many components at different speeds, it can be concluded that the tested system was rather complex in terms of dynamics, which further impeded the unambiguous interpretation of the results.

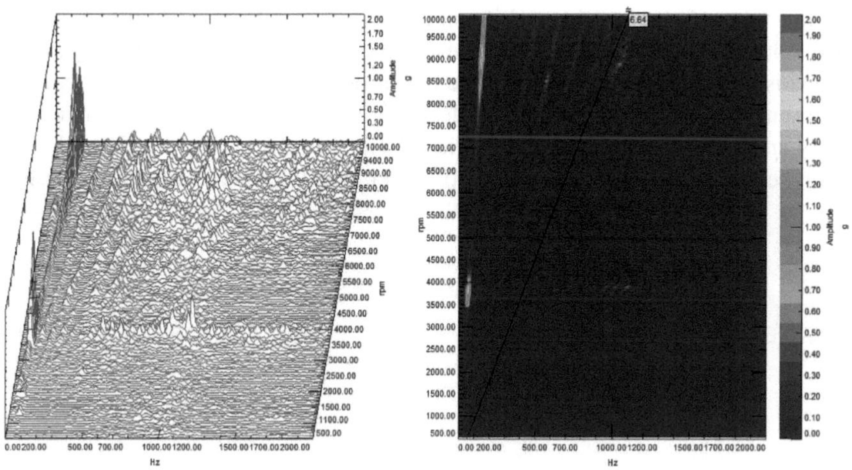

Fig. 5.32 Cascade graph (*left side*) and map of *colours* (*right side*) for a uniaxial acceleration sensor from the second support placed in Y-direction (case of the unbalanced rotor)

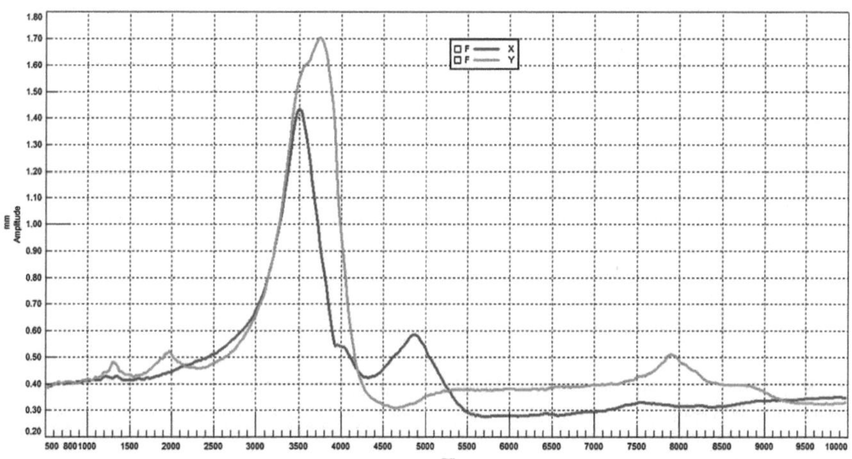

Fig. 5.33 Amplitude of rotor displacements depending on the speed, measured at the disc in the direction of X and Y (case of unbalanced rotor)

The distributions of vibration acceleration recorded on the second bearing support have been presented in Figs. 5.31 and 5.32. They were very similar to the balanced rotor, except that at a speed close to 4,000 rpm an increased level of vibrations in a wide range of frequencies was even more clearly emphasized, which was related to one of the resonances of the system.

The rotor displacements presented in Fig. 5.33 showed that the greatest misalignment of the rotor during rotation in the direction of X (amounting to 1.43 mm at a speed of 3,500 rpm) was over a dozen percent smaller than the displacement in the direction of Y (amounting to 1.7 mm at a speed of 3,950 rpm). Furthermore, the speeds at which the vibration amplitudes were the highest changed only slightly (in comparison to the balanced rotor). The displacements of the rotor after exceeding the resonance speed remained at a low level until the maximum speed was reached.

By analyzing the obtained trajectories of vibrations (Figs. 5.34, 5.35 and 5.36) it can be concluded that in the range of lower speeds (before resonance), despite

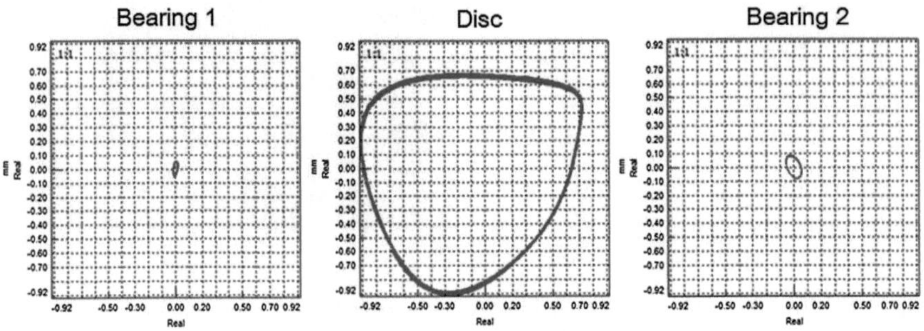

Fig. 5.34 The trajectories of absolute vibrations of an unbalanced rotor with ceramic bearings at a speed of 3,500 rpm

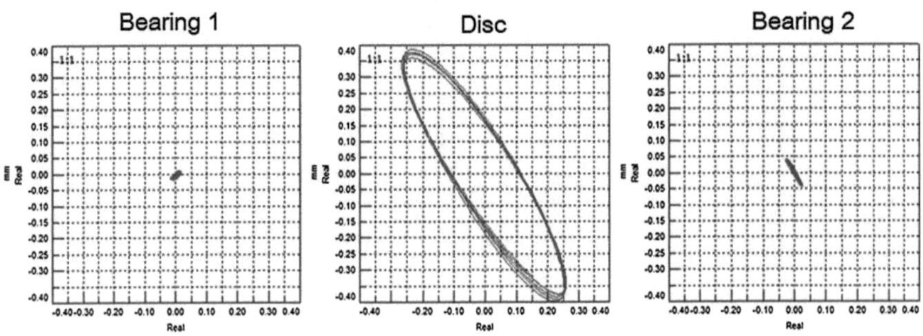

Fig. 5.35 The trajectories of absolute vibrations of an unbalanced rotor with ceramic bearings at a speed of 4,000 rpm

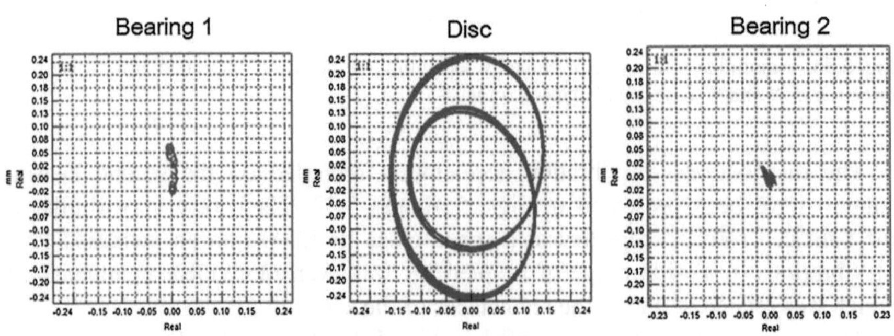

Fig. 5.36 The trajectories of absolute vibrations of an unbalanced rotor with ceramic bearings at a speed of 8,000 rpm

mounting extra unbalanced mass, trajectories of vibrations that were of a similar size were observed. At a resonant speed and above, the trajectories of vibrations changed their shape and the amplitude of vibrations decreased slightly. The beneficial effect of reducing the level of vibrations, along with the increase in speed, was probably due to the effect of self-centering, which proved to be more effective after adding a small mass on the disc. Again, at the highest speeds there were irregular vibrations of the rotor, particularly near the bearing supports.

On the basis of the recorded vibration trajectories, it can be concluded that rotor with ceramic bearings in the prevailing speed range was characterized by similar dynamic properties, as in the case of steel and hybrid rolling bearings. However, at a resonant speed the vibration amplitude increased, which can be seen in Fig. 5.24. It was also noticed that in the case of these bearings, the vibrations of the rotor passed onto the supports to a greater degree, which was visible primarily at higher speeds (Figs. 5.27 and 5.36).

Very interesting is also the comparison of the characteristics of the rotor with ceramic bearings with other, more commonly used bearing systems (steel and hybrid bearings). All the material varieties of rolling bearings enable the safe operation of the rotating system up to a speed of 10,000 rpm. Studies at higher speeds were abandoned due to the limitations of the driving system (clutch) and safety; this did not result from the limitations of the tested bearings. The example set of results obtained for each of the tested systems has been outlined in Figs. 5.37 and 5.38. The following characteristics include the courses of shaft displacements near the disc, recorded in two perpendicular directions for balanced rotors and with a given unbalance. The waveforms presented demonstrate that the properties of the rotor with different variations of rolling bearings were subject to change. At a similar level of unbalance, along with changing steel bearings for hybrid and ceramic ones, the level of the system's vibrations increased at resonant speed, which meant

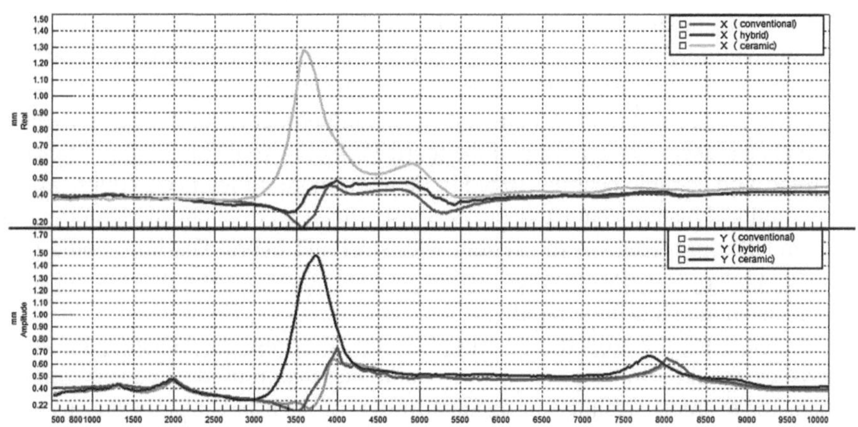

Fig. 5.37 Comparison of displacements for all types of bearings in the direction of X and Y for the balanced rotor (*light blue* and *brown line* ceramic bearings; *blue* and *pink line* hybrid bearings; *red* and *green line* steel bearings)

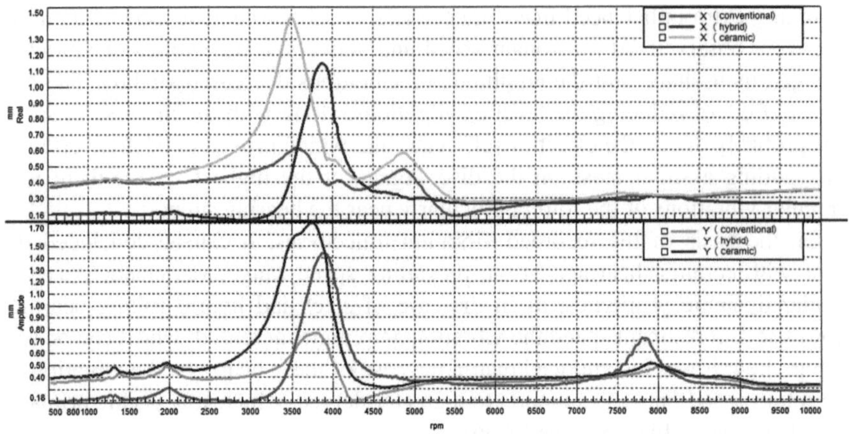

Fig. 5.38 Comparison of displacements for all types of bearings in the direction of X and Y for the unbalanced rotor (*light blue* and *brown line* ceramic bearings; *blue* and *pink line* hybrid bearings; *red* and *green line* steel bearings)

the deterioration of dynamic properties. Changes in resonant speed of the tested systems were also observed. Similar overall conclusions arose after a comparative analysis of other characteristics (cascade graphs, maps of colours, courses of vibration acceleration and trajectories).

In discussing the results obtained, it should be stressed that the studies were performed at room temperature. In order to fully assess the suitability of ceramic bearings for use in microturbines, it would be useful to carry out similar studies also at higher temperatures. Furthermore, it should also be taken into account that ceramic bearings, in contrast to other bearings, were not lubricated with anything, which, among others, could have resulted in their worse damping properties. Only in conditions similar to those that occur when the microturbines are running, would the ceramic bearings demonstrate their advantage over other constructions. High temperature would in fact affect the changing properties of plastic grease and nominal clearances; the leaching of the lubricant factor would also occur. Probably the rotor with ceramic bearings tested in these far more difficult conditions, would have similar dynamic characteristics to those which were obtained at room temperature. The properties of the system with steel and hybrid bearings would presumably undergo significant deterioration, and the bearings themselves could wear out or sustain damage quite rapidly.

By comparing the properties of various rolling bearings, the technological characteristics should also be mentioned. Ceramic bearings proved to be very fragile, which, for example, resulted in damage to one of such bearings during installation on the shaft. Therefore, they require more practical experience and precision of the technical staff. As research work has shown, ceramic bearings also have larger nominal clearances between the rolling elements and rings, which resulted in, among others, a higher level of vibrations. The results obtained and experience gained, however, allow to conclude that ceramic rolling bearings can be quite an attractive and relatively cheap alternative to other bearing systems used in steam microturbines, especially those that operate at lower rotational speeds and in which slide bearings were abandoned for some reason.

5.3 Vibrations of the Radial Microturbine

This subsection discusses the experimental studies of the dynamic properties of the radial microturbine. Since the construction of this turbine has already been discussed in detail in Sects. 4.1 and 5.1, this part contains only information about the test rig in which it was tested, and the method of conducting measurements. The assembly method of the microturbine in the rig has been presented in Fig. 5.39.

The test rig of the microturbine during tests was connected to the rig with heat exchangers (Fig. 5.40) on which phase transformations of the low-boiling medium

Fig. 5.39 Radial microturbine mounted on a test rig

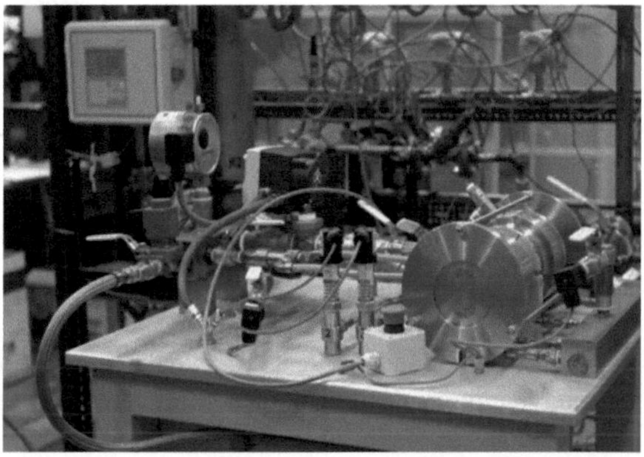

Fig. 5.40 The view from the *top* on the test rig of the microturbine connected to an ORC system [17]

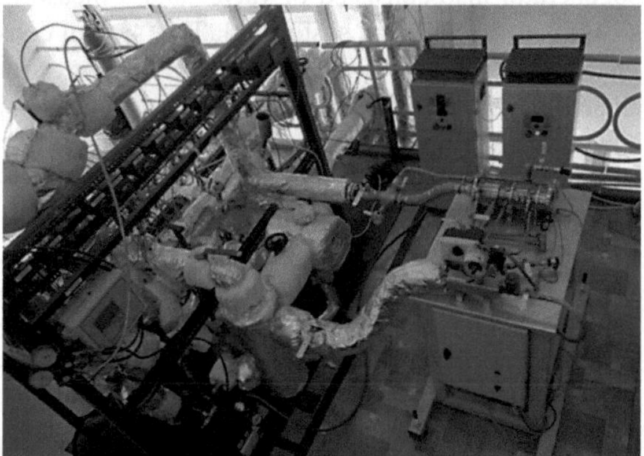

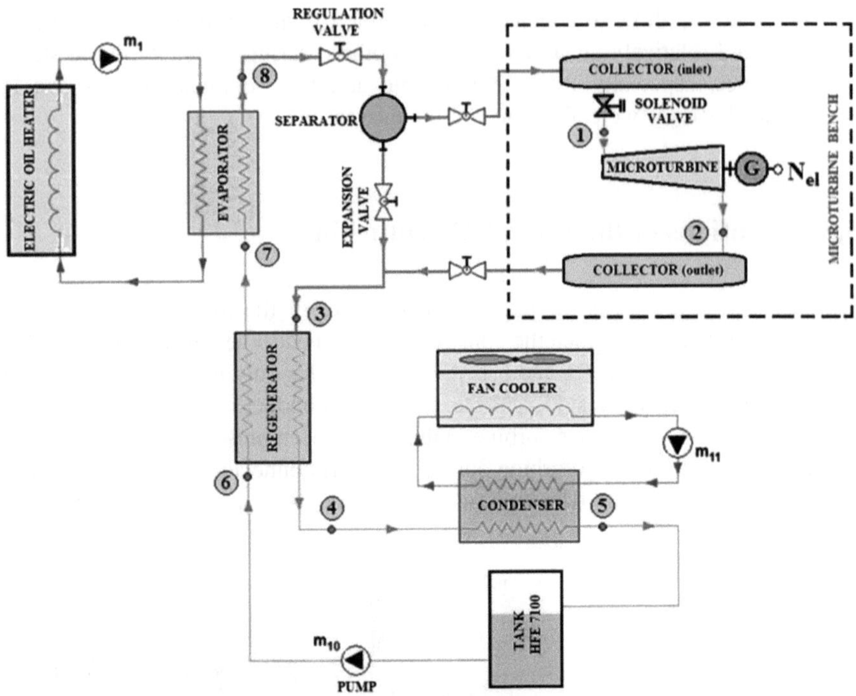

Fig. 5.41 Scheme of ORC system with the microturbine [17]

powering the microturbine were realized. The main objective of the studies conducted in this manner was to determine the thermal and energy characteristics of the microturbine, and the tests of dynamic properties were complementary. The thermodynamic cycle scheme according to which tests of the microturbine were performed has been placed in Fig. 5.41.

The correct functioning of the microturbine's bench, which aside from the turbogenerator also consisted of valves and a measuring system, was supervised by an automatic regulation system [6]. This system consisted also of the conditioning system of electricity produced in the generator of the microturbine. Regulation of all settings of the rig took place remotely from the operator's screen, displayed on a PC monitor (Fig. 5.42). The system also allowed to archive measurement results obtained at the rig. An example characteristic of speed and power generated by the microturbine has been presented in Fig. 5.43.

Before starting the microturbine at the target low-boiling medium HFE-7100, it was checked in terms of dynamic properties when being powered by compressed air. The dynamic characteristics obtained then did not raise any objections. The rotor supported on aerostatic gas bearings worked steadily throughout the expected range of rotational speeds, and the recorded characteristic sizes of vibrations were at a very low level. During the first attempts to start with a low-boiling factor, the very good dynamic properties of the machine were confirmed.

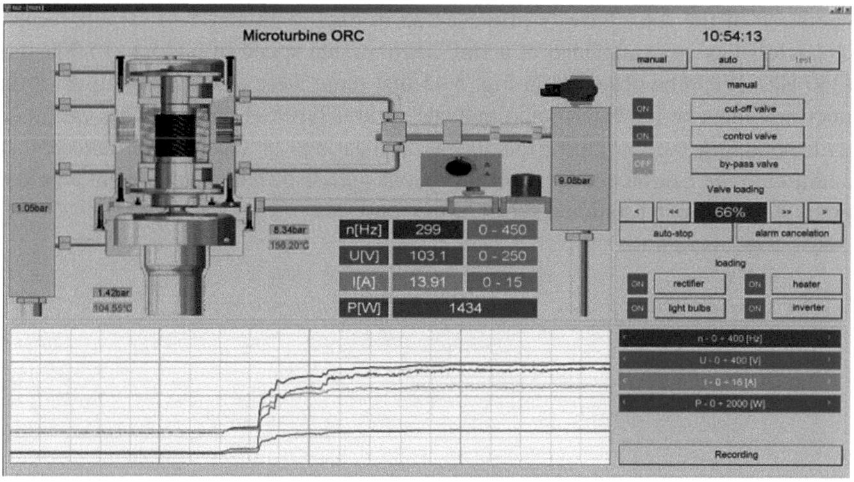

Fig. 5.42 View of the desktop of the operator of the automatic system for regulating the rig of the radial microturbine [18]

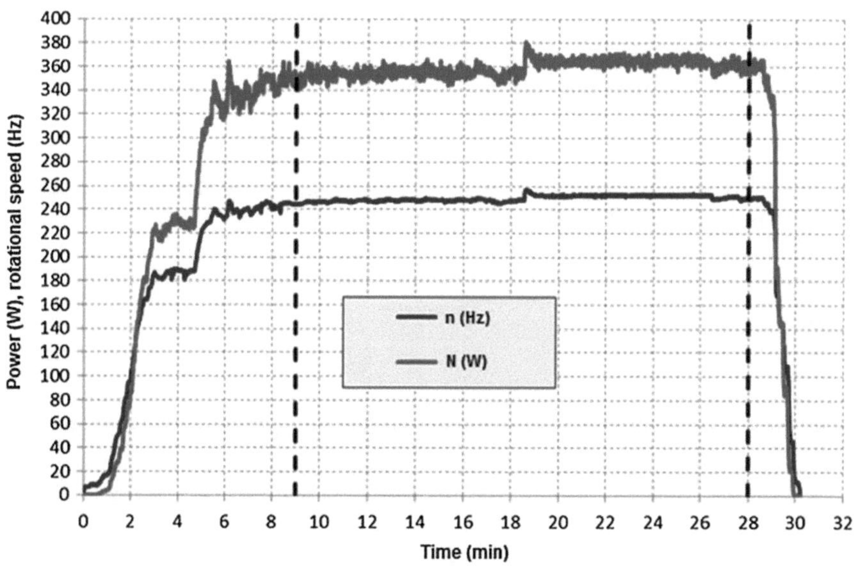

Fig. 5.43 Example characteristics of the microturbine depicting the rotational speed and generated power [17]

Further studies of this machine were conducted in several-week intervals. During one of the subsequent tests, at increasing the rotational speed of the microturbine, disturbing sounds started to occur, and the distributions of vibrations recorded on the body showed the incorrect functioning of the rotating system. The spectra of vibration speeds measured before and during the occurrence of unstable operation of the rotor have been presented in Figs. 5.44 and 5.45. The presented characteristics were obtained at a similar rotational speed of approx. 15,000 rpm (250 Hz). It can be observed in Fig. 5.45 that there were chaotic vibrations of the microturbine's body, which could indicate various types of damages, e.g. problems with obtaining the carrying capacity of the bearings or mechanical damages. In addition to the heightened levels of vibrations and disturbing sound, there was also gradual decrease in rotational speed of the rotor.

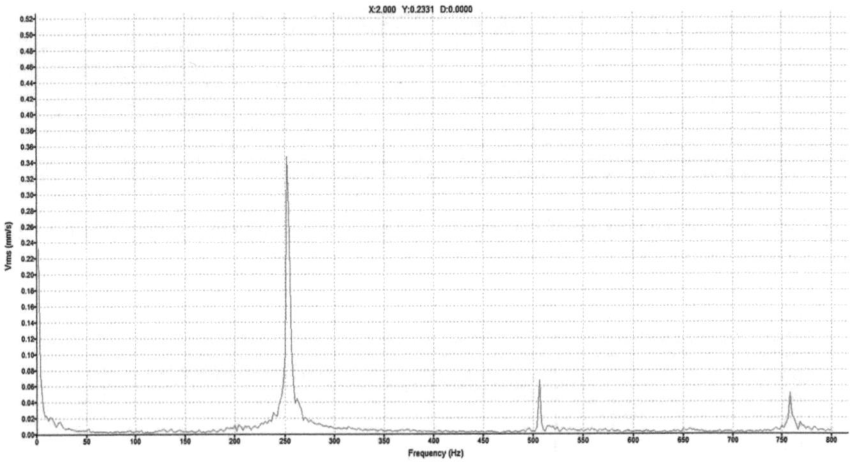

Fig. 5.44 The frequency spectrum of vibration speeds recorded on the body within the bearing before unstable operation of the microturbine occurs [19]

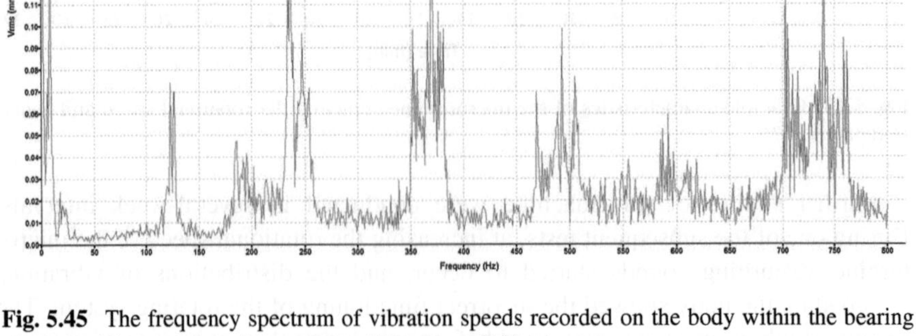

Fig. 5.45 The frequency spectrum of vibration speeds recorded on the body within the bearing after unstable operation of the microturbine occurs [19]

As a result of the problems discussed, further testing of the microturbine was aborted, as its operation in such a state was threatened by the occurrence of failure. After reviewing the results of the tests, it was decided to withdraw the microturbine and carry out a review including checking the rotating elements of the turbogenerator.

After dismantling the airtight body and inspecting the parts of the microturbine, it turned out that inside the machine there were impurities (Figs. 5.46, 5.47, 5.48

Fig. 5.46 The disc of the thrust bearing of the microturbine after disassembly of the back cover of the body [7]

Fig. 5.47 Disassembled rotor wheel (1st and 2nd stage of the turbine) [7]

Fig. 5.48 Disassembled microturbine rotor [7]

Fig. 5.49 Damaged thrust bearing of the microturbine [7]

and 5.49). They caused the formation of sediment on the rotating elements and damage to one of the longitudinal bearings.

The technical condition of microturbine elements after disassembly allowed to place the following diagnostic hypotheses [7]. During subsequent modernizations of the rig with heat exchangers, to which the microturbine rig was constantly connected, various kinds of impurities from the exchangers, pumps and installation elements got into the body. This was made possible through the valves, which during modernization operations were closed, but turned out to be untight. Consequently, for several months, the microturbine rotor was immersed in a mixture of a liquid low-boiling medium, and various types of impurities in the form of: machine oil, welding scale, as well as plastic and metal filings. This resulted in the creation of sediments and clogging parts of the powering holes of the bearing. This is how the rotor became unbalanced and how the bearings lost carrying capacity, which appeared as problems with obtaining stable performance of the rotating system.

In the light of the above, it was necessary to clear part of the turbogenerator, regeneration of bearings and rebalance the whole rotating assembly [7]. After performing these steps, elements of the turbogenerator were again assembled and after a positive check on compressed air it was forwarded for further tests with the use of steam of a low-boiling medium.

The measurements of vibrations on the body of the microturbine within one of the bearings showed that following the regeneration of the rotor and bearings, this machine operated steadily. Figures 5.50, 5.51 and 5.52 present the recorded frequency spectra of vibration speeds for three speeds 18,000, 20,500 and 21,000 rpm. In each of these cases in the vibration spectrum there was a stripe observed, corresponding to the current speed of the rotor. Even at the highest recorded rotational speed the amplitude of the main component of vibration speeds (RMS) did not exceed the value of 1 m/s. The slightly elevated levels of vibrations at low frequencies was associated with the vibrations of the supporting construction and external drives, beyond the test benches. This had nothing to do with the operation of the microturbine.

In conclusion, it can be said that the tested microturbine was characterized by very good dynamic properties. The final results of the tests obtained showed that

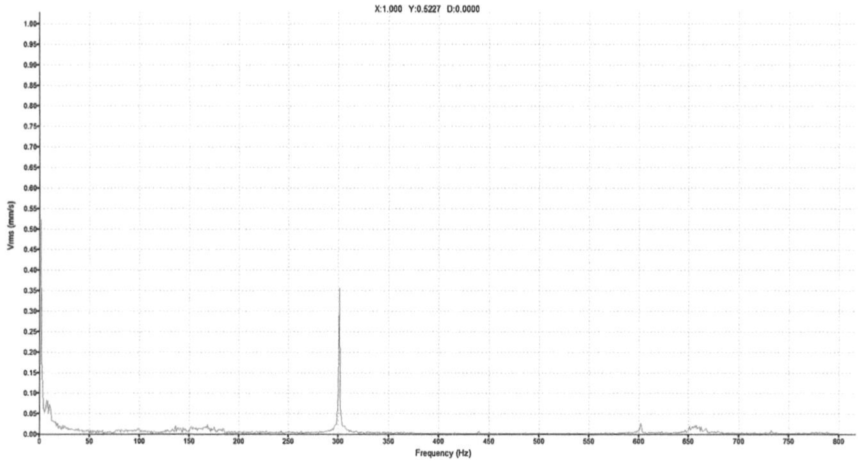

Fig. 5.50 Frequency spectrum of vibration speeds recorded on the microturbine body at a speed of 18,000 rpm

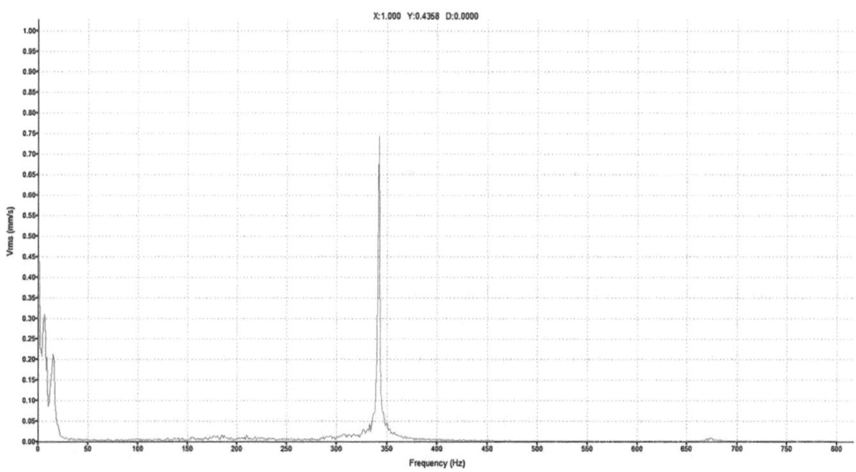

Fig. 5.51 Frequency spectrum of vibration speeds recorded on the microturbine body at a speed of 20,500 rpm

the machine, following regeneration, performed steadily throughout the entire tested range of rotational speeds. The recorded level of vibrations did not exceed the values anticipated by the norms for new machines, freshly commissioned for service. The frequency analysis of the vibration signal also did not reveal any signs of any defects. Further studies have also shown that, in addition to very good dynamic properties, the designed and built radial microturbine had the thermal-flow and energy characteristics similar to those in the design. Therefore, in the

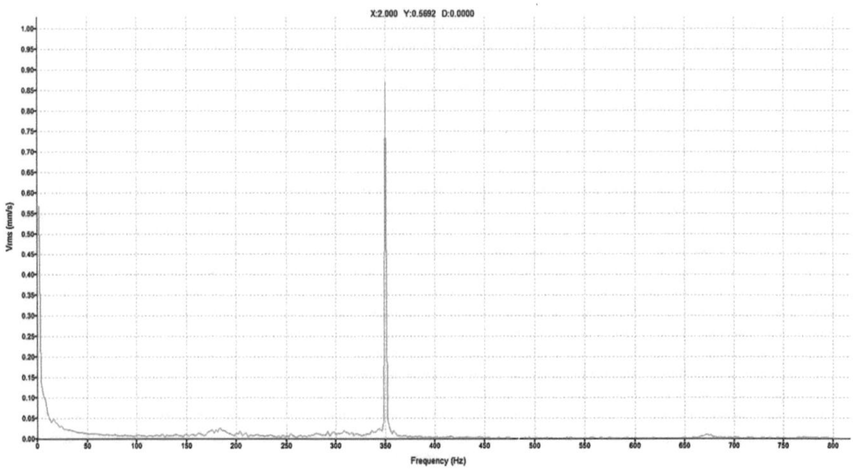

Fig. 5.52 Frequency spectrum of vibration speeds recorded on the microturbine body at a speed of 21,000 rpm

prototype of the ORC micro power plant [8] being developed under the project it was decided to apply a twin solution of the microturbine, with some modifications concerning only the fluid-flow system.

5.4 Vibrations of an Ultrasonic Microturbine

The ultrasonic microturbine was created within the framework of the realized project based on the previously discussed and well tested construction of the radial microturbine. Instead of four stages of blades (two centripetal and two centrifugal), there was one centrifugal stage applied in it (Fig. 5.53), in which the speed of the flowing steam was greater than the speed of sound [9, 10]. The rotor of this microturbine also had a greater nominal rotational speed of approx. 35,000 rpm. Thanks to the solutions applied, this machine was also characterized by a higher intrinsic efficiency.

Initially it was expected that at a higher rotational speed it will be possible to use aerodynamic foil bearings on the rotor. After solving the problems related

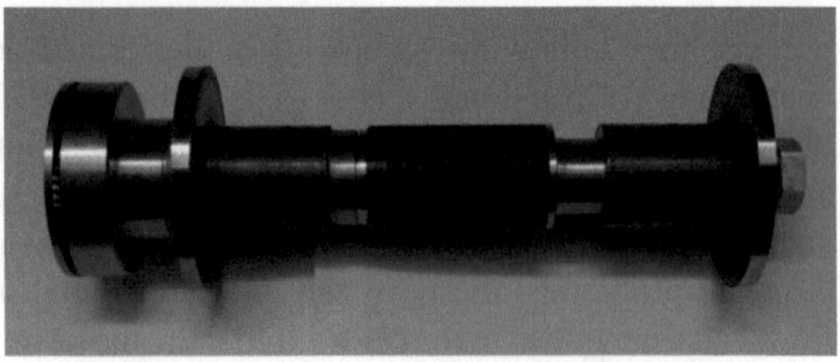

Fig. 5.53 Assembly of a rotating ultrasonic microturbine [10]

to the durability of foil bearings, it turned out, however, that they require a large start-up moment, and during the operation of the rotor, the bearing journals can move even up to 0.3 mm. These factors caused that in the ultrasonic microturbine, where it was difficult to obtain a large start-up moment and very small blade clearances were applied, innovative foil bearings were abandoned and replaced with previously tested aerostatic bearings. Therefore, the tested machine had the same bearings as the previously discussed radial microturbine [11].

The purpose of the tests discussed in this subsection was the performance of characteristic vibrations measurements and the evaluation of the dynamic state of the ultrasonic microturbine during operation with the target operational factor of the ORC micro power plant. Due to the innovative solution of the fluid-flow part, the unconventional bearing system (gas bearings lubricated by steam of a low-boiling medium), as well as a high rotational speed, the results of the tests may be very interesting for people dealing with the dynamics of microturbines and are not available in literature.

Tests of the ultrasonic microturbine were conducted on an integrated micro power plant test rig. It was built as the first integrated version of the ORC system, in which all the subassemblies were mounted on one supporting structure and were controlled and monitored using one system of automatic regulation [12]. Long distances and easy access to all subassemblies of the system made it possible to exchange them quickly and introduce the necessary modernizations. The rig used during the tests is therefore a universal solution, allowing, among others, the performance of tests of various structural solutions of microturbines, pumps, heat exchangers, valves, etc. The tested ultrasonic microturbine during tests was at the highest accessible location of the rig (Fig. 5.54), which secured it against flooding by the operating medium in liquid form. The main task of the test rig, in addition to playing the role of a supporting structure, was powering the tested microturbine with steam of a low-boiling medium of the appropriate parameters, and then the collection of used steam.

The tested ultrasonic microturbine was developed as an airtight construction, as in order to power the bearings, steam of the same operating medium as for powering a fluid-flow system was used [9, 11]. The basic elements of the microturbine were: single-stage radial turbine, electric generator with so-called rare earth magnets and sets of two transverse and longitudinal bearings. All the subassemblies were enclosed in an airtight body with a water jacket cooling the generator. The microturbine was characterized by the following design parameters:

- nominal rotational speed—35,000 rpm;
- internal power of the turbine—3.26 kW;
- initial pressure—1.1 bar;
- initial temperature—156 °C.

The external appearance of a microturbine and method of mounting an ORC micro power plant on a test rig has been presented in Figs. 5.54, 5.55 and 5.56. Figure 5.56 also shows the location of the vibration sensor (accelerometer) used during the measurements. Due to the fact that the body of the microturbine was

Fig. 5.54 The test rig of the ORC micro power plant with an ultrasonic microturbine mounted in the upper part [13]

Fig. 5.55 The ultrasonic microturbine on the test rig of the ORC micro power plant—view from the side of the collector powering the bearing [13]

made from austenitic stainless steel, in order to mount the sensor on the magnetic base, it was necessary to install an additional threaded element made of carbon steel. In this microturbine, because of the air tightness of the body, it was not possible to directly measure the vibrations of the shaft or disc. During the studies

Fig. 5.56 The ultrasonic microturbine on the test rig of the ORC micro power plant—visible place of mounting the accelerometer [13]

conducted, the rotational speed of the turbine reached up to 25,300 rpm, at an electrical power slightly exceeding 1,800 W.

During the tests, the mobile vibration analyzer Diamond 401AXT was used with a uniaxial accelerometer, mounted by using a magnetic base, along with dedicated software for archiving, visualizing, processing and evaluating the measurement results. The analyzer used allowed to keep different types of measurements, facilitating the diagnosis and evaluation of the dynamic state of various types of machines.

Within the experimental studies of the dynamic properties of an ultrasonic microturbine measurements and records were made of the following:

- time courses of vibration amplitudes;
- frequency spectra of vibration speeds;
- the effective value of vibration speed, the effective value of vibration amplitude, and peak-to-peak amplitudes (for the assessment of the dynamic state).

The most interesting, selected results of the tests have been presented in the diagrams below. All the recorded characteristics of an ultrasonic microturbine can be found in the work [13].

5.4.1 Time Courses

Figures 5.57 and 5.58 present the time courses of vibration displacements, recorded in the body of a microturbine. Recording takes place in a time of 16 s, without any additional processing of the signal. Therefore, these are the results of measurements in a so-called raw form, very difficult to interpret, but which enable further processing and analysis. In using time courses it is particularly difficult to interpret high-frequency vibrations, therefore, these issues will be discussed further in the next part of this subsection.

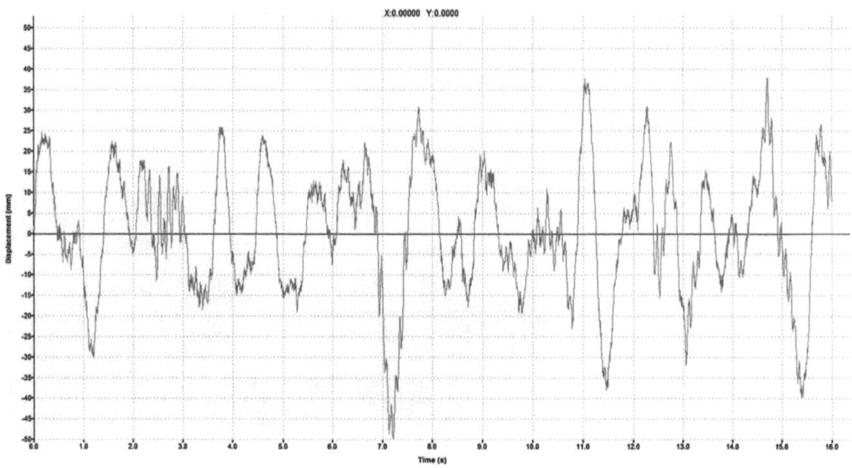

Fig. 5.57 The time course of vibration displacements recorded in the body at a speed of approx. 21,000 rpm (example 1)

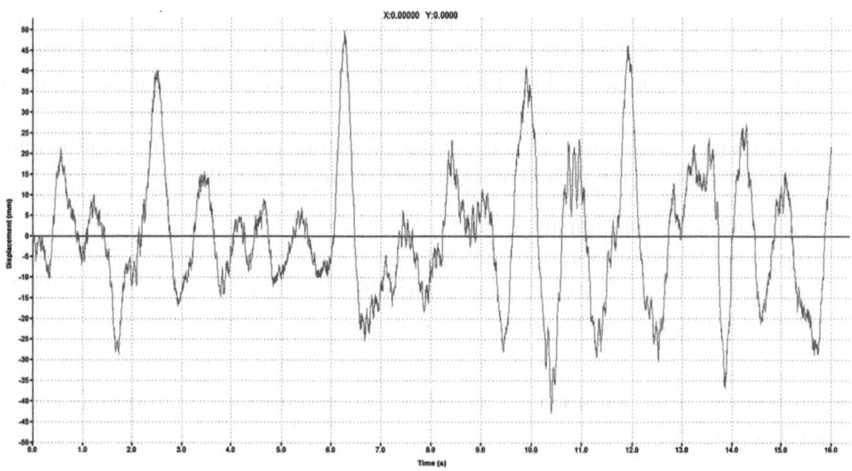

Fig. 5.58 The time course of vibration displacements recorded in the body at a speed of approx. 21,000 rpm (example 2)

A characteristic feature of recorded time courses was high-amplitude vibrations at a low frequency. The amplitude of these vibrations reached up to 50 μm. The drawings presented show that these were non-periodic vibrations. Due to the low frequency it may be assumed that these vibrations triggered external factors and they were further strengthened as a result of the application of a fairly flaccid frame construction, on which the microturbine was mounted. Therefore, these vibrations were not associated with the work of the turbogenerator itself.

5.4.2 Frequency Analysis

Measurements of the vibration spectrum were performed at the following analyzer settings:

- range of frequencies—800 Hz;
- resolution of measurement—800 lines;
- averaging—RMS;
- number of averagings—3;
- window—rectangular.

Due to the wide range of rotational speeds of the microturbine, the measured size was the speed of vibrations, expressed in mm/s. A few dozen spectra of vibrations were recorded at different rotational speeds of the microturbine. Due to the large volume of all the results, Figs. 5.59, 5.60, 5.61, 5.62, 5.63 and 5.64 show the selected vibration spectra for different rotational speeds. In accordance with the growing numbering of drawings, the rotational speed gradually increased, and then decreased. Due to the long-lasting and dependent on a few mutually coupled factors, the process of obtaining the desired parameters of steam of the low-boiling medium, and the irregular course of increasing rotational speed of the microturbine associated with it, in this case the typical measurements of the start-up and shut-down of this machine were abandoned.

On the basis of the measurement results obtained, it can be concluded that the obtained distributions of vibration speeds are of high quality and clear. Throughout the entire measured range of rotational speeds, three components of vibrations occurring dominated: at very low frequencies up to over a dozen Hz, at frequencies in the range of 50–60 Hz, and at higher frequencies.

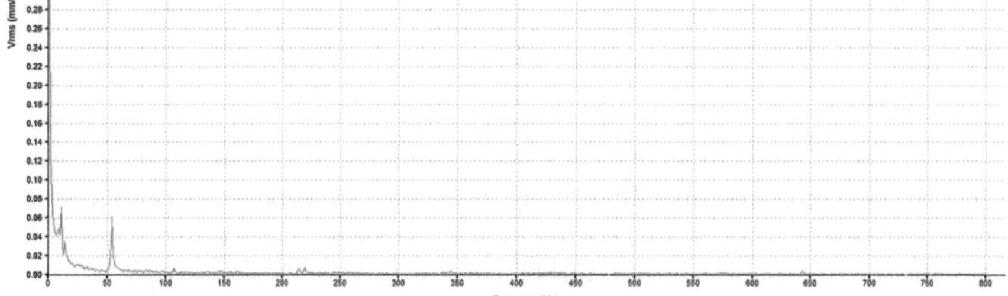

Fig. 5.59 The frequency spectrum of vibration speed measured on the body of the microturbine before its start-up with a visible excitation of vibrations from the operating circulation pump

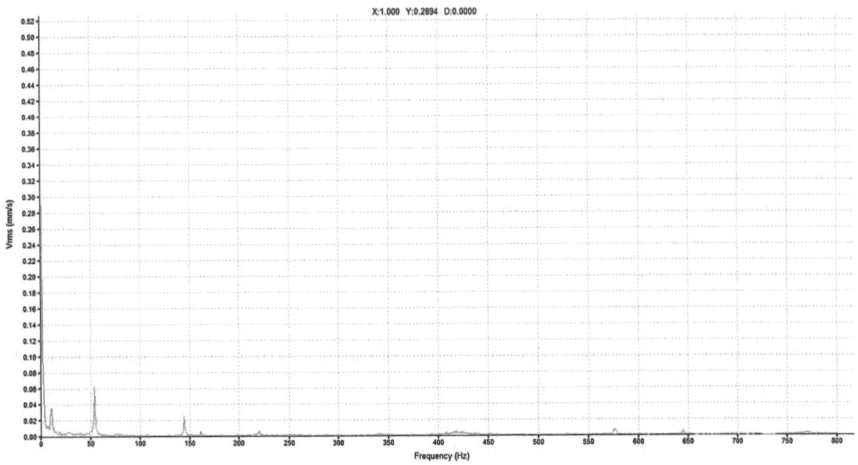

Fig. 5.60 Frequency spectrum of vibration speeds measured at the speed of the microturbine rotor of approx. 8,600 rpm

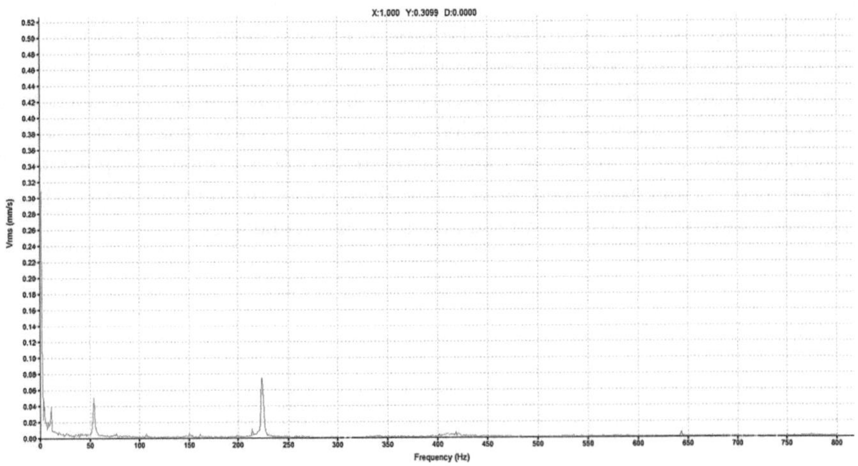

Fig. 5.61 Frequency spectrum of vibration speeds measured at the speed of the microturbine rotor of approx. 12,800 rpm

In the lowest range of frequencies fairly large amplitudes of vibration speeds could be observed, but depending on the next measurement they occurred in the range from a few to over a dozen Hz. They were associated primarily with the vibrations of the whole construction of the test rig, which was driven by external factors, e.g. walking on the platform on which the rig was situated, switching valves, opening the control boxes, the pulsating factor in pipelines, etc. Therefore, these were not frequencies directly related to the work of the microturbogenerator.

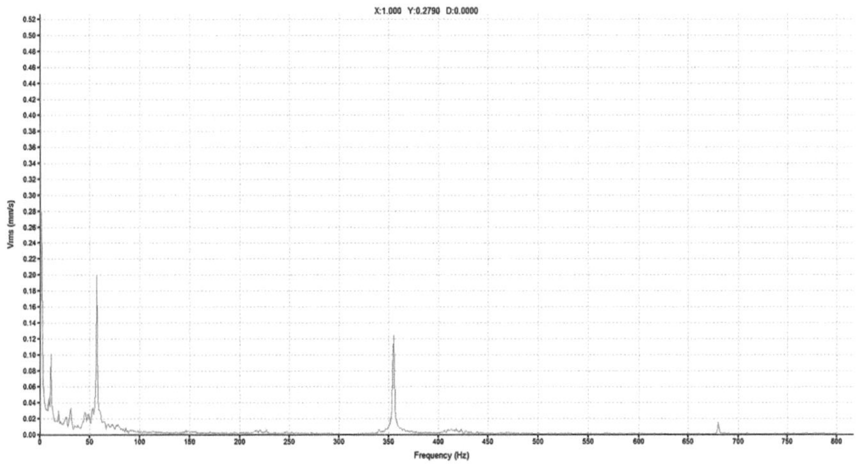

Fig. 5.62 Frequency spectrum of vibration speeds measured at the speed of the microturbine rotor of approx. 21,300 rpm

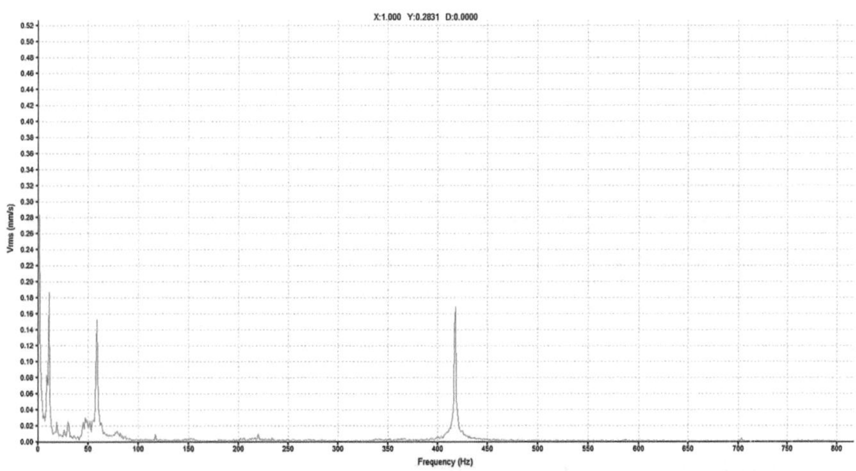

Fig. 5.63 Frequency spectrum of vibration speeds measured at the speed of the microturbine rotor of approx. 25,100 rpm

On all the presented spectra of vibrations, an increase in the vibration amplitude was recorded in the range of 50–60 Hz (amplitude reached up to 0.2 mm/s). These vibrations were associated with the work of the circulation pump, which was situated in the lower part of the rig. The recorded frequencies corresponded to the rotational frequency of the pump's rotor. The rather frail frame construction caused the displacement of these vibrations on the entire rig.

Increased amplitudes of vibrations visible in the drawings above at higher frequencies (from 60 Hz) corresponded to the rotational frequencies of the

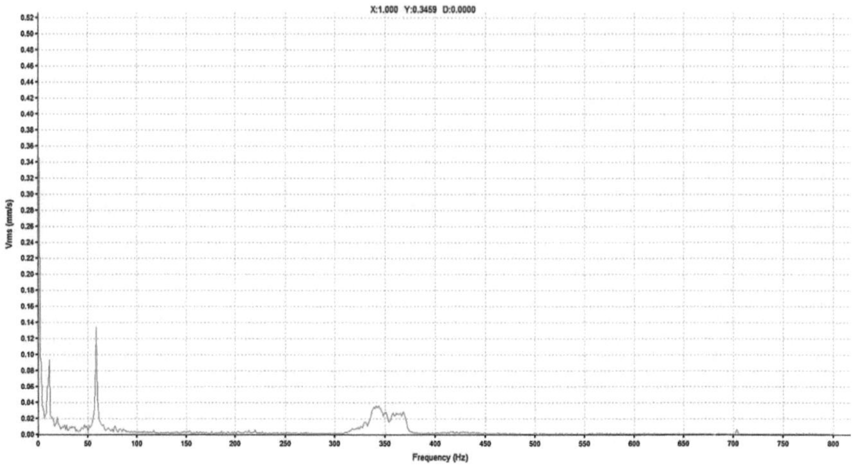

Fig. 5.64 Frequency spectrum of vibration speeds measured at the speed of the microturbine rotor of approx. 20,600 rpm

microturbine rotor (synchronous vibrations). Starting from Fig. 5.60 an acceleration of the rotor followed, until the speed exceeded 25,000 rpm, which can be seen in Fig. 5.63. Then there was a decrease in rotational speed. Depending on whether the measurement was made at a fixed rotational speed, or during its change, the stripe corresponding to the rotational speed was clear ("sharp") or fuzzy (like in Fig. 5.64). Aside from the synchronous stripe, no other component vibrations were observed, which could result from the improper operation of the microturbine. The maximum recorded amplitude of vibration speed of the body, throughout the speed range did not exceed 0.18 mm/s. In comparison with the amplitude of the pump's vibrations, which was far away from the location of measurement, it was a very small value.

5.4.3 *Evaluation of the Dynamic State*

The characteristic size of vibrations applied to evaluate the dynamic state of high-speed rotating machines is most often the effective value of vibration speeds VRMS measured in mm/s. The standards used to evaluate the dynamic state of machines, as the reference value use the effective value of vibration speeds [14]. The measurement of this parameter of vibrations is done on the non-rotating elements of the machine, usually on the bearing housing. In the case of the discussed tests, the measurement was performed on the microturbine body, near one of the bearings.

The measurements of effective vibration speeds were performed with the same mobile device—the vibration analyzer Diamond 401AXT with an attached uniaxial accelerometer. During the measurement, a filter with a frequency of 10–1,000 Hz [14] was used, at an averaging time equal to 3 s. The measuring range of the apparatus used amounted to 0–500 mm/s for the vibration speeds. The vibrations analyzer used also allowed the measurement of effective values of the vibration amplitude and the peak-to-peak amplitude.

The measurement was conducted at two operational states of the microturbine:

- State I—rotational speed approx. 21,300 rpm, power approx. 1,600 W;
- State II—rotational speed approx. 23,700 rpm, power approx. 1,800 W.

The measurement for each operational state was performed several times at intervals, and the final results were indicated as average values. The obtained values have been presented in Table 5.1.

In terms of power, according to the standard [14], the tested microturbine belonged to group I. For this group of machines, the threshold V_{RMS} values, allowing to qualify the machine's state to one of the stages are as follows: A/B—0.71 mm/s; B/C—1.8 mm/s; C/D—4.5 mm/s. While stage A indicates the vibrations of a machine newly commissioned for service, and stage D—vibrations that may indicate a damaged machine. The measured values allow to qualify the tested microturbine to stage A.

To summarize the test results of the ultrasonic microturbine, it can be concluded that it was characterized by very good dynamic properties. Throughout the studied range of rotational speeds, the level of vibrations recorded on the body of this machine was very low, and the amplitude of the vibration speeds corresponding to the synchronous component did not exceed 0.18 mm/s. Moreover, in the analyzed range of rotational speeds, no worrying over- and sub-synchronous components appeared, which also indicated the stable operation of the rotor-bearings system. This was also confirmed by the measurements of characteristic sizes of vibrations, including the values of effective vibration speed. All the measured values appeared with large reserves in the zone expected for machines newly brought into service. Such results were obtained despite the fact that in the recorded vibration signal on the microturbine body vibrations deriving from the circulation pump were dominant, which was the source of vibrations of the entire testing environment, and thus also of the tested microturbine. It can be concluded that the studies carried out have confirmed that the prototype ultrasonic microturbine, in terms of dynamic properties, was designed and constructed properly. The applied system of two transverse-longitudinal gas bearings, lubricated by steam of the low-boiling factor ensured the proper transfer of operational loads and excellent dynamic properties.

Table 5.1 The results of the measurements of characteristic sizes of vibrations

Marking of state	V_{RMS} (mm/s)	A_{RMS} (μm)	$A_{P\text{-}P}$ (μm)
I	0.133	0.934	5.124
II	0.146	0.886	5.018

5.5 Vibrations of the Microturbine Supporting Structure

Because the dynamic properties of the microturbine supporting structure may affect their vibrations, these constructions also require appropriate analysis. This is particularly important in the case of new, yet untested constructions, where an improperly placed microturbine can transfer vibrations onto the entire device, or other subassemblies can have a negative effect on the operation of the turbogenerator.

Because the carrying construction of the micro power plant test rig, built in the laboratory of IMP PAN, was made as a lightweight mesh structure, it was necessary to check its dynamic properties [15]. Proof that it is excellent in transferring vibrations of the rig's subassemblies could be the vibration spectra presented in the previous subsection, where in the measurement made on the microturbine body vibration components can be seen, induced by the operation of the pump (located at a great distance from the microturbine). In the case of the bearing construction, it was found that it is necessary to determine the modes and frequencies of free vibration in the range of frequencies of operation of the bench's subassemblies, which will enable the introduction of any structural changes, or the use of appropriate techniques of vibration insulation.

The tested construction was built from aluminium profiles with a side of 40 mm (Figs. 5.54 and 5.65). Individual elements were screwed together with connectors, forming the system of profiles used. During the studies, the whole construction was placed on a ground made of wooden planks covered with plastic lining. The weight of the construction was approx. 65 kg, and the dimensions of the base 66 × 66 cm at a height of 213 cm.

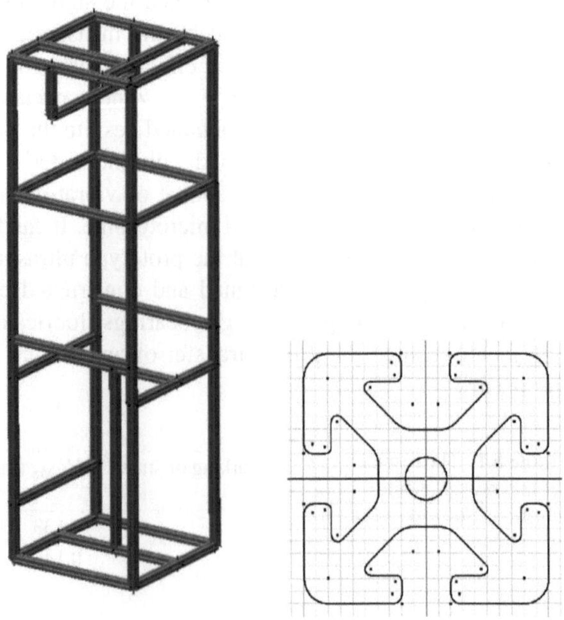

Fig. 5.65 The model of the supporting construction of the test rig and cross-section of the aluminium profile

5.5.1 Method of Testing

For inducing vibrations in the construction, two electromagnetic inductors were used from the company TIRA (models TV 50018 and TV S 51110), which were attached to the construction using elastic connectors. At the end of the connector there was an impedance head installed, which was attached to the profiles using a two-ingredient resin. The use of a magnetic mounting base was not possible, as the construction was made of aluminium. Due to the necessity to constantly change their position, the accelerometers were fitted using wax.

The control unit and data acquisition module was the analyzer SCADAS Mobile from the company LMS. The used analyzer had 16 inputs, as measuring channels and two inputs for tachometers and two outputs, e.g. for controlling vibration inductors with amplifiers. During the measurements, two impedance heads were used (PCB 288D01). Each of them took up 2 measurement tracks, as it measured force and acceleration at the same time. Additionally, two triaxial accelerometers were used (PCB M356B08). The wires connecting the sensors with the analyzer were shielded. SCADAS Mobile was connected through an Ethernet interface with a personal computer with the installed software Test.Lab.

The first stage of work was to prepare in the Test.Lab program the measurement geometry consisting of measuring points (Fig. 5.66). The location of these

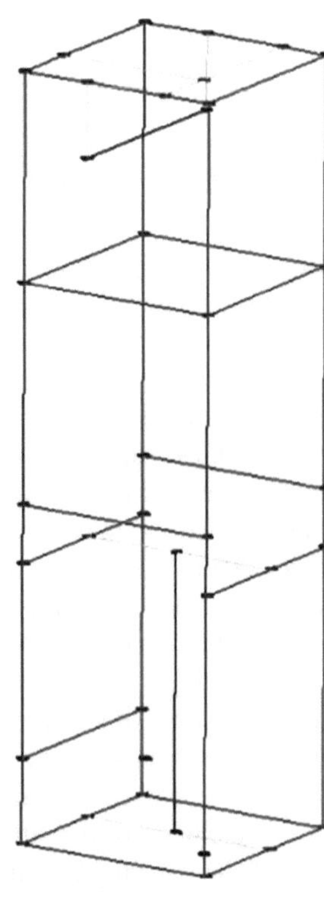

Fig. 5.66 Measurement geometry developed in the Test.Lab program [15]

points was selected so that the measuring point fell in place of connecting two or three profiles with each other. For auxiliary purposes, before carrying out an experimental modal analysis, a simplified numerical model of the tested system was developed and a simulation modal analysis was conducted. Thanks to this, already at the stage of preparing for experimental tests, the points in the construction could be indicated in which, at specific modes, nodes and arrows of vibrations occurred.

A total of 36 measuring points were selected, which should ensure sufficient measurement accuracy. Every increase in the number of points would result in an extension of testing time and complicating geometry—and consequently, every irregularity in the construction of the frame (e.g. a loose profile) would entail difficulties in obtaining the correct results of the measurement. The tests were performed with the following settings:

- bandwidth—1,024 Hz;
- spectral line—2,048;
- resolution—0.5 Hz;
- data acquisition time—2 s.

The drive had the nature of a white noise with a set driving level of 0.5 V. A single measurement consisted in fitting two sensors on the object, causing driving and observation of coherence between the driving and response of the system. Only if the coherence of signals was deemed to be correct (driving energy was replaced with vibrations of the object) did a change in the place of fitting the sensors follow.

5.5.2 Test Results

Because already during the performed tests it was noticed, that the dynamic properties of the tested construction were constantly changing, it was decided to carry out the entire study twice. This aimed to determine the level of changes occurring and indicate the source of the emerging problems. Based on the results obtained, stabilization diagrams were drawn up for the two independent measurements. As a result of the fact that very many maximums occurred in the entire range of frequencies, which could indicate the presence of free vibrations (Fig. 5.67), this range was divided into several zones so as to facilitate the identification of stable poles due to: frequency, damping coefficient and modal vector. The modes of free vibration selected in this way were compared with one another using the Modal Assurance Criterion (MAC) [16], which has been presented in Fig. 5.68.

According to the results presented in Fig. 5.68, it can be concluded that during the tests, some of the free vibration modes of the system somewhat changed, especially in the range of lower frequencies. Changes in how often the same modes of vibration occurred, reaching a maximum of 3 Hz, were also noticed.

On the basis of the studies conducted, the cause of changes of dynamic properties of the carrying construction of the test rig was found. It turned out that, as a

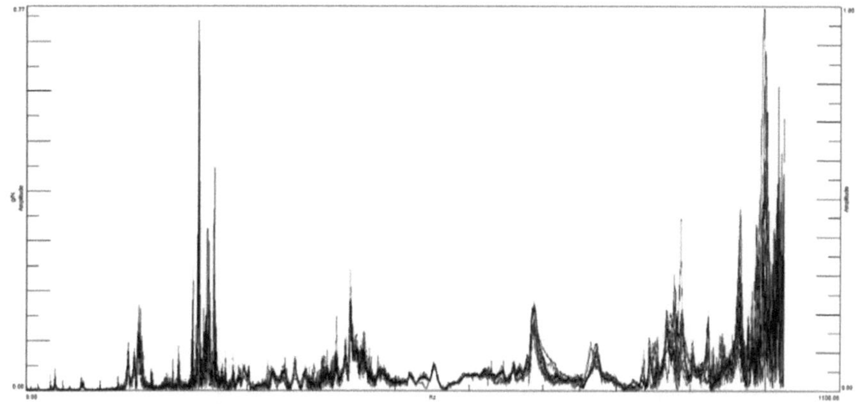

Fig. 5.67 The frequency spectrum of dynamic responses of the construction in driving points for all series of measurements

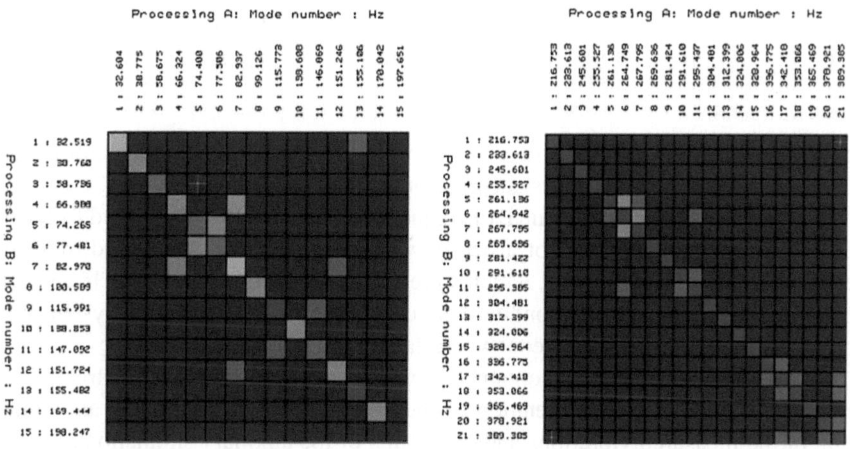

Fig. 5.68 Comparing 36 modes of free vibration of the carrying construction obtained in two measurements using the MAC criterion

result of inducing vibrations with electromagnetic generators, the threaded joints used in standard connections supplied with the profiles loosened. These were original connectors provided by the manufacturer of the profiles, however, as it turned out, they were not resistant to the effects of dynamic loads. Even very strong screwing of these elements did not guarantee certainty of connection, and too much tension in the screw caused the local deformation of the aluminium profiles. In the light of the above, it was difficult to perform repetitive measurements, and a very simple and quite effective solution was to apply additional resilient pads, not provided in the standard system of fitting profiles. Such a solution did not

Table 5.2 A summary of the experimentally determined frequencies and damping coefficients for 30 free vibration modes

No. of mode	Frequency (Hz)	Damping (%)	No. of mode	Frequency (Hz)	Damping (%)
1	32.52	0.52	16	216.75	0.04
2	38.76	0.38	17	233.61	0.09
3	58.74	0.47	18	245.60	0.19
4	66.39	0.73	19	255.53	0.11
5	74.27	0.78	20	261.14	0.18
6	77.48	0.69	21	264.94	0.17
7	82.97	0.41	22	267.80	0.19
8	99.13	0.66	23	269.64	0.06
9	115.99	0.31	24	281.42	0.15
10	138.85	0.42	25	291.61	0.16
11	147.92	0.42	26	295.39	0.43
12	151.72	0.43	27	304.48	0.25
13	155.48	0.73	28	312.40	0.28
14	170.04	0.12	29	324.01	0.09
15	198.25	0.06	30	328.96	0.3

guarantee total resistance of the connection to the effects of vibrations, however, the loosening of the connectors occurred on a much smaller scale.

Throughout the analyzed range of frequencies, close to 100 modes and frequencies of free vibration of the construction were indicated. The 30 selected frequencies and damping coefficients, which were set at the lowest frequencies have been compiled in Table 5.2. The condition concerning low damping in the system was met, because at the selected frequencies, damping did not exceed 1 %.

On the basis of the results obtained, a visualization of the modes of free vibration was created. The interpretation of these results was facilitated by an animation of the measured vibrations and the results of the auxiliary simulation analysis performed using the MES model. Because the experimentally determined free vibration modes of the assembled constructions are usually not very clear, for convenience, they have been presented below together with the corresponding results of the simulation modal analysis.

Figure 5.69 shows the vibration mode occurring at approx. 82 Hz. These were bending vibrations of the entire construction. The next Fig. 5.70 presents torsional vibrations at a frequency close to 100 Hz. The next mode of free vibration (Fig. 5.71) is no longer as easy to specify, as mainly the longest vertical profiles took part in the vibrations, and the lower part of the frame almost did not move. In the following cases, very good compatibility of the experiment was obtained with the results of the computer simulation, which was not achieved for other modes of vibration, the frequencies of which have been provided in Table 5.2. The compatibility of the modes were observed usually in relation to individual fragments of the system. This was due, above all, to the fairly complex geometry, the very large

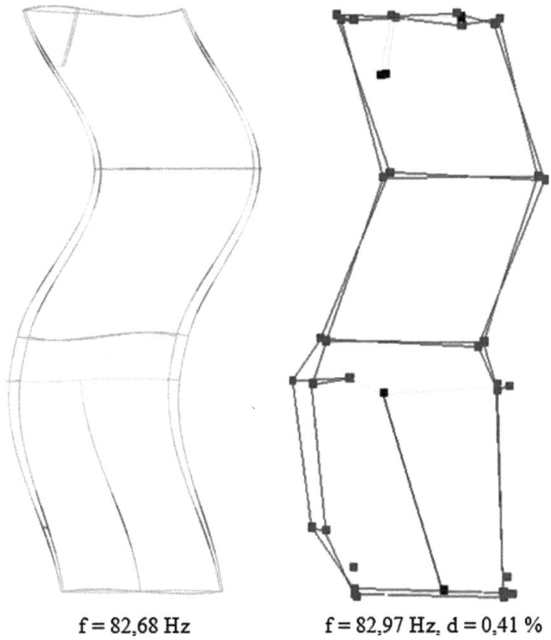

Fig. 5.69 The mode of free vibration determined by simulation and experimentally of the test rig frame occurring at approx. 83 Hz

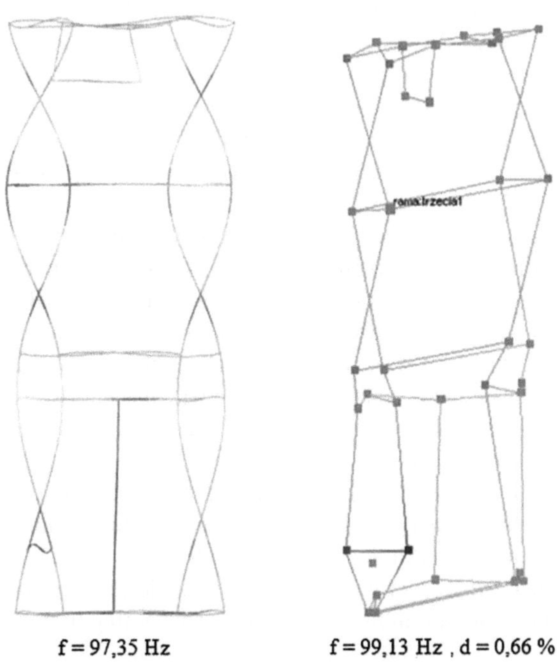

Fig. 5.70 The mode of free vibration determined by simulation and experimentally of the test rig frame occurring at approx. 99 Hz

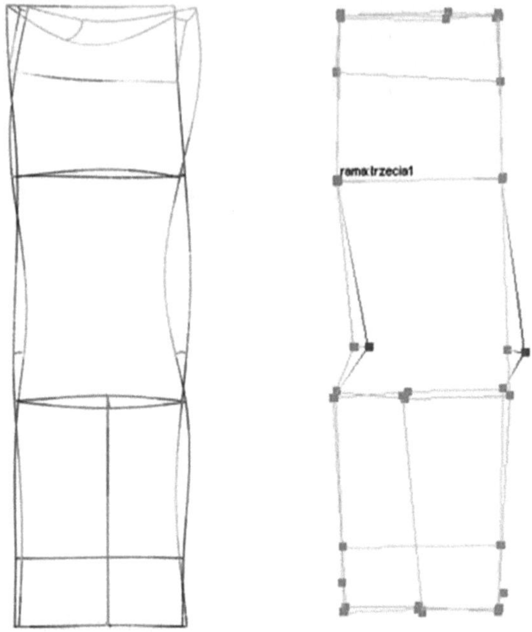

f = 173,16 Hz f = 170,04 Hz, d = 0,20 %

Fig. 5.71 The mode of free vibration determined by simulation and experimentally of the test rig frame occurring at approx. 170 Hz

number of frequencies of free vibrations, and the previously mentioned problems with obtaining fixed and unchanging in time connections between the profiles.

Generally, as a result of experimental modal analysis, 80 modes and frequencies of free vibration were obtained. By way of computational analysis, as many as 151 modes were determined, which resulted mainly from the fact that the numerical analysis allows one to additionally indicate strongly dampened modes of vibration, which are difficult to measure with the applied methods.

The presented example of a modal analysis of the supporting construction of the microturbine demonstrated how complex the dynamic characteristics of a system, with which such machines work, can be. The discussed studies were conducted only for the supporting frame itself, which greatly facilitated the measurements and analysis of the results obtained. In fact, all subassemblies of a device are fitted on a construction, such as heat exchangers, pumps, valves, control and measurement equipment, as well as the entire installation. These subassemblies significantly change the modal model of the system and make it even more difficult to analyze.

Regardless of the complexity of the construction and degree of difficulty of dynamic analysis of the supporting systems of microturbines, such studies should be performed in modern, high-tech devices. In addition to it being possible to evaluate the impact of support on dynamic characteristics of microturbines, they can facilitate solving many problems concerning high-speed rotating systems.

References

1. Kozanecki, Z., Więckowski, S., Łagodziński, J., Klonowicz, P., Borkowski, R.: Preliminary draft and technical documentation of the turbo-generator prototype with aerostatic bearings. Int. Study IMP PAN, No. Arch. 137/2010, Gdańsk-Łódź, in Polish (2010)
2. Rządkowski, R., Drewczyński, M., Kiciński, J., Żywica, G.: Calculation tests of the turbo-generator's rotor variants selected for the further implementation—modal analysis. Int. Study IMP PAN, No. Arch. 113/2010, Gdańsk, in Polish (2010)
3. Łuczak, M., Żywica, G., Bagiński, P.: Experimental modal analysis of the radial microturbine's rotor. Int. Study IMP PAN, No. Arch. 100/2011, Gdańsk, in Polish (2011)
4. Bagiński, P., Żywica, G., Kiciński, J.: Experimental studies of the dynamic properties of the microrotor with ceramic rolling bearings. Int. Study IMP PAN, No. Arch. 431/2013, Gdańsk, in Polish (2013)
5. LMS International: LMS Test.Lab User Manual. Rev. 12, Leuven (2012)
6. Kaczmarek, A., Kozanecka, D., Hanausek, P., Kozanecki, Z., Miazga, K., Gizelska, M.: Implementation of the start-up and operation monitoring system for the turbogenerator prototype under electric energy consumption conditions—installation of the system, software integration and initial testing. Int. Study IMP PAN, No. Arch. 411/10, Gdańsk, in Polish (2010)
7. Kozanecki, Z., Więckowski, S., Łagodziński, J., Marczyk, A., Tkacz, E., Ihnatowicz, E., Żywica, G., Niewiadomski, J., Bykuć, S.: Assessment of the causes of failure, regeneration, balancing and qualification start-up tests of second prototype of the ORC turbogenerator. Int. Study IMP PAN, No. Arch. 441/2013, Gdańsk, in Polish (2013)
8. Feasibility Studies of the Project: Model agroenergy complexes as an example of distributed cogeneration based on local renewable energy sources. IMP PAN Gdańsk, in Polish (2008)
9. Kozanecki, Z., Klonowicz, P., Więckowski, S., Łagodziński, J., Miazga, K.: Conception, calculations and technical documentation of alternative radial turbines for the third turbogenerator prototype functioning with model ORC power plant. Int. Study IMP PAN, No. Arch. 22/2012, Gdańsk, in Polish (2012)
10. Kozanecki, Z., Marczyk, A., Więckowski, S., Łagodziński, J., Borkowski R.: Technical documentation, technological analysis and making of ready frame elements and semi-finished products (for CNC machining) for use in the third prototype of supersonic turbine's turbogenerator and assembly of the machine on the micro power plant test rig with ORC. Int. Study IMP PAN, No. Arch. 142/2012, Gdańsk, in Polish (2012)
11. Kozanecki, Z., Marczyk, A., Więckowski, S., Łagodziński, J., Borkowski, R.: Design and implementation of aerostatic gas bearing being a component of third turbo-generator prototype with supersonic turbine and machine tests using air as a working medium (with electricity generation measurement. Int. Study IMP PAN, No. Arch. 632/2012, Gdańsk, in Polish (2012)
12. Breńkacz, Ł., Bykuć, S., Niewiadomski, J.: Design of the micro power plant test rig containing a high-speed supersonic turbine. 2D, 3D figures. List of additional materials required for assembly and connection of the turbogenerator. Int. Study IMP PAN, No. Arch. 844/2012, Gdańsk, in Polish (2012)
13. Kiciński, J., Żywica, G., Banaszek, S.: Experimental studies of the dynamic properties and the evaluation of the dynamic state of supersonic microturbine operating with target working medium. Int. Study IMP PAN, No. Arch. 985/2013, Gdańsk, in Polish (2013)
14. ISO 10816-1:1995: Mechanical vibration—Evaluation of machine vibration by measurement on non-rotating parts—General guidelines
15. Bagiński, P.: Modal analysis of the test rig's frame bearer of the micro CHP power plant with ORC. Laboratory testing and simulation studies. Int. Study IMP PAN, No. Arch. 324/2012, Gdańsk, in Polish (2012)
16. Palacz, M., Krawczuk, M.: Vibration parameters for damage detection in structures. J. Sound Vib. **249**(5), 999–1010 (2002)

17. Kaczmarczyk, T., Ihnatowicz, E., Żywica, G., Bykuć, S., Kiciński, J., Kaniecki, M., Janicki, W.: Start-up tests of the radial microturbine using the test rig for testing heat exchangers as a source of low-boiling medium's vapour. Determination of characteristics of the microturbine and ORC cycle. Int. Study IMP PAN, No. Arch. 889/2012, Gdańsk, in Polish (2012)
18. Kiciński, J., Żywica, G., Bykuć, S., Kaczmarczyk, T., Ihnatowicz, E., Kozanecki, Z., Kaczmarek, A.: Experimental studies of the radial microturbine with target working medium after the regeneration of the rotor and bearing system. Int. Study IMP PAN, No. Arch. 473/2013, Gdańsk, in Polish (2013)
19. Kaczmarczyk, T., Ihnatowicz, E., Kiciński, J., Żywica, G., Bykuć, S.: Start-up tests and determination of the characteristics of radial microturbogenerator on the modernized test rig for testing heat exchangers in the ORC installation. Int. Study IMP PAN, No. Arch. 398/2013, Gdańsk, in Polish (2013)

Index

[1] *Note*: Page numbers followed by "f" and "t" indicate figures and tables respectively.

J. Kiciński and G. Żywica, *Steam Microturbines in Distributed Cogeneration*,
DOI 10.1007/978-3-319-12018-8

MIX
Papier aus verantwortungsvollen Quellen
Paper from responsible sources
FSC® C105338

If you have any concerns about our products,
you can contact us on
ProductSafety@springernature.com

In case Publisher is established outside the EU,
the EU authorized representative is:
Springer Nature Customer Service Center GmbH
Europaplatz 3, 69115 Heidelberg, Germany

Printed by Libri Plureos GmbH
in Hamburg, Germany